ECOLOGICAL FANTASIES

ECOLOGICAL FANTASIES

DEATH FROM FALLING WATERMELONS

Cy A. Adler

A DEFENSE OF INNOVATION, science and
rational approaches to environmental problems.

GREEN EAGLE PRESS
99 Nassau St.,
New York, N. Y. 10038

GREEN EAGLE PRESS

99 NASSAU STREET
NEW YORK, N. Y. 10038

Printed in the United States of America.
Library of Congress Catalog Card Number 73-80095

ISBN # 0-914 018-02-7

Much of the discussion on "Limits to Growth" in Chapter 16 first appeared as a book review in the Village Voice on 25 May, 1972, and is reprinted by permission of the Voice. Part of the section entitled "Falling Watermellons" first appeared in Science on 4 November, 1972.

We are grateful to Liveright Company for permission to reprint material from Bertrand Russel's "Conquest of Happiness".

The drawing of the archer standing on one leg was scratched on a cave wall in northern Iberia some 30,000 years ago. Most of the other figures were drawn by William Segal. Miscellaneous art was created by Pedav.

CONTENTS

PREFACE

After the Romans withdrew from England, the myths about King Arthur and his gallant knights grew among the Celts, who were forced into the hills by conquering tribes. Current myths about the environment have arisen in a way similar to the myths about King Arthur. Both mythologies were formulated by a fearful people living in troubled times in need of tales of victory over hated and fearsome forces about which they knew little. The stories of wicked giants, goblins spewing fire and brimstone , and monsters, environmental or otherwise, became wonderfully magnified at each retelling by the wandering bards. Each anxious teller embroidered the myths with his own personal concerns. And through this forest of fearful doom, with lofty sounding ideals, galloped the hero knights in shining armor. King Arthur may or may not have existed, but the tales about him make wondrous reading. The modern believer of myths has his own set of shining knights who fight the monstrous pollution makers. Many of these new myths based on a modicum of truth and an overdose of fear make fearful reading.

Many of the knights of the environmental mythology are indeed noble and brave, though quite a few of the gallants appear somewhat flawed in their vision. Who can but admire the late, gentle Rachel Carson, or the bold and ascetic Ralph Nader taking on the dragons of the chemical and automotive industries? Who can help but admire the erudite seer, Barry Commoner, and fail to take notice of that stern minstrel of fairy tales, Paul Ehrlich? However, some of the gallants who have taken up lances to tilt against real or fancied environmental monsters leave much to be desired in the way of integrity and intelligence. Even among the Knights of the Round Table there were a few miscreants.

During the past few years it has been fashionable (and to some, profitable) to predict worldwide environmental disaster. Some of the more rabid environmentalists have been assuring us that the environment is doomed and that mankind does not have long to live on this planet. For instance, the French oceanographer Jacques Cousteau estimates that 40 percent of the world's marine life has disappeared because of industrial pollutants in the sea, and the rest is on its way out. B. Connor gives us "40 years to live," while P. C. Orloff assures us that we have only 15 years to go. Paul Ehrlich is more charitable and predicts only that "In the 1970's hundreds of millions of people are going to starve to death." Ehrlich appears to be saying that there will be about 5.5 to 7 billion people on earth by the end of this century and the starvation of a few hundred million people will still leave a goodly number of souls to carry on the race.

Pollution is very much America's problem because though we have only 6 percent of the world's population, we Americans use about 40 percent of the world's non-renewable natural resources and, with our cars, factories, power plants and suburbs, account for approximately half of the world's industrial pollution. And in America, pollution abatement, like motherhood, is an issue that everyone (almost) is in favor of. Bandwagon politicians have been among the worst offenders in promoting schemes to abate pollution, without thought as to the reality of the problem and the socioeconomic effect of their actions.

Hasty anti-pollution actions forced by environmentalists often result in rash efforts that are counterproductive. For instance, the oil companies, in a rush to remove tetraethyl lead from gasoline, have added other chemicals that may prove more harmful to the environment than lead, and which are increasing the emission of other pollutants to the air. The ban on phosphate-based detergents may have caused more environmental harm than their continued use would have. The discontinuance of DDT has caused insects, some of them carrying disease organisms harmful to man and other creatures, to run rampant in many areas and food production to decline. Opposition to the construction of new power plants in rural areas has led to the inevitable construction of plants in already polluted urban areas, resulting in annoyance and potential harm to a much greater segment of the population.

Indeed there is little question that man can destroy himself and mutilate the earth's ecosystem; but if he does, it is more likely to be the result of political and social folly than because of pollution. Before the era of mass communication each man had his own private fantasies, and myths propagated slowly from individual to individual. But now a lunatic with a microphone and money can spread his version of unreality across the face of the land spilling myths into millions of preconditioned ears. Thus a few men may bring about catastrophe by causing a nation to believe and act out some pernicious political or social fantasy to the point where reason holds no sway.

I can envision a "yellow-peril fantasy" causing a cataclysmic crusade that could easily destroy two billion people. I can imagine an "anti-system, anti-science" fantasy gripping the television-grown youth of the Atlantic community and generating a civil war that would pull America and Western Europe down into another Dark Age. But, except in time of war, I cannot envision modern nations destroying whole environments, such as the Pentagon appears to have done with its herbicide bombings in Southeast Asia.

A beautifully illustrative example of how environmental myths proliferate was inadvertently revealed by Michael Belknap, Director of Mayor Lindsay's Council on the Environment. In an article, "The American Way of Life: Is It Killing Us?" (The New York Times Supplement, April 18, 1971), Mr. Belknap writes, "Although proof of the cause and effect of pollutants and disease is not established, the evidence is overwhelming." Mr. Belknap appears to connect lack of proof with overwhelming evidence, but we shall see that the evidence on air-pollution hazards as well as other environmental issues is far from convincing. One may ask, what is the good of evidence when people are determined to come to conclusions in direct variance with the findings? I have a firm belief, perhaps misguided, in man's ultimate rationality.

7

My aim in writing this book is to inform a lot of intelligent citizens, who I feel are now grossly misinformed regarding the environment, by presenting them with hard facts about pollution and various other ecological hobgoblins, and by drawing logical conclusions from the facts. It is my hope that when the good citizens' anger subsides it will be replaced by thought, and followed by positive, rational actions that will improve our environment.

Since 1962, when Silent Spring ushered in the era of ecological hysteria and concern, the public has been entertained by a series of environmental scenarios predicting doom. Some of the environmental myths have taken the popular fancy and caused a good deal of jumping about and unwarranted fear; and as Edmund Burke observed, "No passion so effectually robs the mind of all its powers of acting and reasoning as fear."

Some of the more persistent environmental myths are: the air in our cities is killing us all; mercury and lead are poisoning us; Lake Erie is dead or at least dying from eutrophication; detergents are ecologically disastrous; DDT, 2, 4, 5-T and PCB's are contaminating nature; radiation and thermal pollution are destroying vast amounts of life; oil spills are permanently killing ocean life; et cetera. I do not accept all of these myths, and hope to help dispel them.

There are many environmental problems that are not treated in this book simply because I do not think they are myths. In some cases not enough is known about them to say anything meaningful. Some of these issues are strip mining; noise; the SST; nitrate concentration of ground water; long-term weather modification; and a host of other specialized problems arising from use of new chemicals and products. Industrial man produces more than one million different kinds of products that eventually wind up as waste. It is obviously impossible to deal with them all.

For the past three years I have been engaged in developing new systems relating to aquaculture, offshore storage and transfer of bulk materials, water-pollution and pathogen indicator systems, large-diameter underwater pipe systems, wave measuring devices, and systems for using waste sludge to build up beach dunes. The money to support

8

The New Yorker
21 August, 1971

*"And grant that I may take into my system only acceptable
levels of mercury, cadmium, lead, and sulphur dioxide."*

these fascinating and potentially useful endeavors has
come entirely from public stockholders and from consulting
on environmental issues. Many of the systems that have
been developed have anti-pollution potential, but this is
not the place to expound on them. In addition to my
industrial work, I have taught at the New School for Social
Research "Science for the Citizen" program, where I
gave courses in marine science and ecological fantasies
and realities.

In part, this book is my search for reality. I have been
working in the fields of oceanography, marine science
and ecology for more than 12 years. During the past few
years I have become uneasy when technical, disinterested
reports crossing my desk conflicted with the stories of
environmental disaster that were filling the press. Was
it true that the world would soon come to a polluted end?
I tried to find out; and this book is an attempt to separate
ecological truth from ecological muddle.

I realize that many people are loath to part with their
visions of catastrophe, and apologize if my findings disturb
their assurance of imminent ecological doom.

9

Cy.Adler

APOLOGIES

In late 1971 I went on a Samuel Johnson tangent for reasons that are obscure. The binge was initiated by a chance quotation; "No, Sir, when a man is tired of London he is tired of life," which seemed apt to a lover of New York wishing to defend his city from ill informed and malicious attacks. In the preface to his dictionary, Johnson writes:

> I saw that one inquiry only gave occasion to another, that book referred to book, that to search was not always to find, and to find was not always to be informed; and that thus to pursue perfection, was, like the first inhabitants of Arcadia, to chase the sun, which, when they had reached the hill where he seemed to rest, was still beheld at the same distance from them.

In writing Ecological Fantasies I feel as Dr. Johnson did, that to cover all aspects of environmental pollution and all the follies perpetrated by self-styled defenders of nature would be like chasing the sun, an endless task. The flood of reports, articles, speeches, meetings, debates, programs and books on environmental subjects is almost overwhelming. There were more than 500 paperback books in print in late 1971 on the subject. Even after limiting my current reading to articles in Science, the New York Times and about eight specialized environmental journals, keeping abreast of the literature on this subject is an arduous task; I could easily rewrite whole chapters of this book on a monthly basis with new material, none of which would alter my basic assumptions. Therefore, like Johnson, I had to set limits to my work so that it would be finished before the world came to an end, and trust that it contains sufficient matter to arouse and please discriminating readers.

DEDICATION

This book is dedicated to all those people who during the past 10 years have been scared unnecessarily by falacious reports of environmental catastrophe.

If I were to choose one individual to whom to dedicate Ecological Fantasies it would be Rachel Carson. I assigned her book The Sea Around Us to my class when I first started teaching oceanography at the State University Maritime College some nine years ago. I was so taken with her style and knowledge that I wrote her a letter suggesting that we work together revising The Sea Around Us as a college text in oceanography. She wrote me saying that though she appreciated the nice things I said about her writing, she did not like working with coauthors. It was Miss Carson's Silent Spring, written in 1962, shortly before she died, that initiated the era of environmental concern among the American middle class and unleashed a flood of unsubstantiated reports of ecological doom and catastrophe.

ACKNOWLEDGMENT

Hundreds of people have helped me with this book; some willingly, many unknowingly by sending me reports and answering seemingly innocent letters and questions. Among the former are Jack Wise, a marine biologist, whose incisive comments and suggestions were most helpful, and Deborah Solomon Wallace, who disagreed with practically everything that I wrote; George Claus, an Hungarian micro-biologist, with whom I started to write, but who parted company with me along the way; Elsie Wood, city planner; Connie Simon, meteorologist; Jacques Wolfner and Paul Gilbert, sanitary engineers; Sam Atkin, psychiatrist; Max Tohline, engineer; Elsie Zumwalt, classics scholar and teacher; Scotty Grumet, businessman; Hank Baslow, biologist; Milt Comanor, materials specialist and engineer; Len Walit, insurance specialist; Lloyd McAulay, patent attorney; James Friend, Eric Posmentier and Jerome Spar, geophysicists; Edward Marston, physicist; Glenn Paulson, chemist; Tom Burke, lawyer; Mark Leymaster, law student; students from my class in environmental fantasies at the New School for Social Research and many others. And of course my wife, Pat Adler, who read and corrected drafts till ecology flowed out of her ears; and some dozen typists who corrected many mistakes. But mistakes undoubtedly remain that are mine and no one else's.

PART ONE: POLLUTION AND ECOLOGICAL FANTASIES IN A COMPLEX SOCIETY

Even when the facts are available, most people seem to prefer the legend and refuse to believe the truth when it in anyway dislodges the myth.

John Mason Brown: <u>Saturday Review</u>

1

CLEARING THE AIR

"Nature is a mutable cloud which is always and never the same."
Ralph Waldo Emerson

FROM GOD, NO COMMENT

The inherent plan of nature appears to be to allow each earthly species to utilize or abuse the existing environment in its own favor. In Edmund Burke's words, "All progress is based upon a universal innate desire on the part of every organism to live beyond its income." In the process one species may help or depend on another, forming part of an ecological cycle; or a species may work to the detriment of, or even extinguish another species. So be it. The Creator looks down on the scramble for survival of the little creatures below and says nothing. He does not comment when the beaver builds dams that destroy the breeding grounds of fish and inundate the burrows of surrounding animals. He says nothing when ants build mounds and tunnels that kill all competing species in the area. He does nothing (as far as I know) when the mightiest builders of them all, the coral-building marine polyps, create vast accretions of coral islands and atolls, thus preempting the region for its successors and friends. In the Great Barrier Reefs the polyps have built a structure of greater proportions and volume than all the combined constructions and excavations of all the humans who have ever crawled over this earth. So be it.

Every species, including man, has a well-defined physical and biological environment or "niche" in which it can live. The oyster requires brackish water between 32° and 68° F.; the chamois prefers cold, craggy mountains; anerobic bacteria require an environment devoid of oxygen, and man requires an atmosphere with oxygen, in which he can keep his body temperature at 98.6° F. In nature, almost all species except man have adapted to specific niches. A complex habitat, such as a tropical forest, provides more niches and supports a greater variety of species than does a simpler habitat such as a desert. When a habitat is in equilibrium, all or most of the available niches are occupied. As the habitat changes, the less adaptable species become extinct and their former niches are occupied by newly evolved or better adapted species.

Most species adapt either by specialization, in order to fit a particular niche more closely, or by generalization, to fit a wider variety of niches; that is, by broadening their adaptability rather than by simplifying it. Man is the most conspicuously adaptive organism, although the rat, the pigeon, the cockroach and the sparrow have adapted well to city life.

Man is fundamentally different from the other animals of the earth in that he has an innate drive to humanize the world. Through millions of years of evolutionary development animals have attempted to adapt to the earth in a genetically simple manner to survive and propagate; man modifies his environment in order to express his humanity and in so doing he has at times destroyed the natural systems that sustain him and other forms of life.

DEFINITIONS

What do the terms "ecology," "environment" and "pollution" mean in a technical sense? It is questionable whether the commonly accepted definitions of these terms are adequate to serve as a basis for serious study.

Ecosystems are the basic communities of nature. An ecosystem is a natural grouping of nutrients, minerals, atmospheric gasses, plants and animals and their organic debris all linked by the flow of food, nutrients and energy from one part of the system to another part. The living and non-living parts of the ecosystem interact with and affect each other. Atoms of oxygen and nitrogen are absorbed by animals that produce proteins and carbon dioxide. The carbon dioxide in the air is absorbed by green plants in the presence of light to produce carbohydrates that are eaten by the animals in the ecosystem.

Animals, plants and nutrients are essential for the maintenance of any ecosystem. In each stable ecosystem, inputs and outputs must balance.

By environment we mean the aggregate of all the external conditions and influences affecting the life and the development of an organism. The biosphere in which life can survive consists of air, water and land, and covers the planet earth as a thin shell. (Figure 1) Each of these three components of the environment are subject to pollution and ultimate sterility.

The Committee on Pollution of the National Research Council, National Academy of Sciences, defines pollution as:

• • • an undesirable change in the physical, chemical, or biological characteristics of our air, land, and water that may or will harmfully affect human life or that of other desirable species, our industrial processes, living conditions, and cultural assets; or that may or will waste or deteriorate our raw material resource. (1968)

This broad definition has been narrowed somewhat by the California State Water Control Board in its definition of water pollution: "Pollution of water is any impairment of its qualities that adversely and unreasonably affects its subsequent beneficial use." In this definition change must be both adverse and unreasonable to be termed "pollution." The belief that a change from the natural state is not necessarily adverse but may, on the contrary, be beneficial, is implicit in the definition. Furthermore, the definition also implies that any substance introduced into the water might be considered as a potential pollutant if its concentration is large enough to affect the subsequent beneficial uses of the water.

Both definitions of pollution quoted above center on man's use of the environment. Although in the first definition reference is made to "other desirable species" the wording of the statement indicates that the desirability of change is evaluated primarily according to its effect on the human species. In a broad sense all of man's activities contribute to some degree to the pollution of the environment. Biological existence necessarily leads to the accumulation and elimination of waste

17

STRATOSPHERE — UP TO 25 MILES ABOVE THE EARTH

TROPOSPHERE
5 TO 10 MILES HIGH

Mt. Everest –
5.5 miles high

BIOSPHERE
THE THIN LAYER GIRDING
THE EARTH IN WHICH LIFE
SURVIVES

LITHOSPHERE
THE SOLID EARTH

TO THE CENTER OF THE
EARTH – 4,000 MILES

HYDROSPHERE
OCEANS, LAKES, WATER VAPOR.
MOST OCEAN LIFE LIVES IN
UPPER 600 FT. 71% OF
EARTH'S SURFACE.

DEEPEST TRENCH – 6.4 MILES IN DEPTH.

Fig. 1-1

A cross section of the planet Earth. From the surface the solid earth extends about 4000 miles to the liquid center of the earth; and the atmosphere extends upward to 25 miles above the surface. The BIOSPHERE, that relatively thin shell in which life as we know it, exists, is only about 12 miles thick at most. A few hardy animals live in the deep ocean trenches six miles below the surface and a scattering of microbes wander about the troposphere. Most life hugs the surface of the solid earth or is concentrated in the upper 600 feet of the ocean.

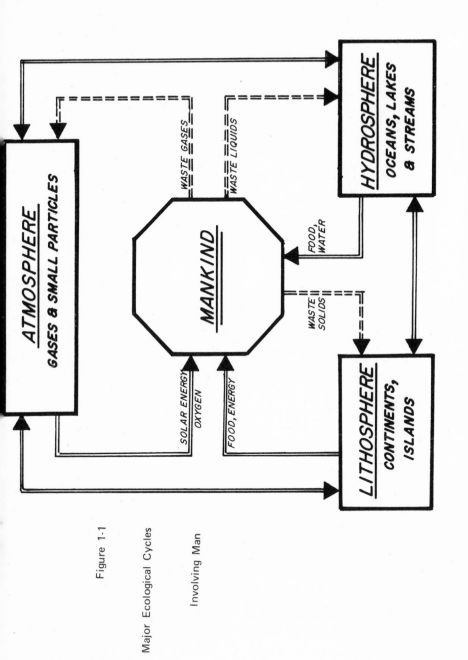

Figure 1-1

Major Ecological Cycles

Involving Man

products, and all of these products entering the environment may be considered pollutants. Thus, to define precisely what constitutes environmental pollution becomes an almost impossible task. Fresh water entering estuaries is a serious pollutant from the standpoint of the oyster drill and, to a lesser estent, the oyster. Salt water intrusions into loose sedimentary strata and the subsequent replacement of underground fresh-water reservoirs are forms of natural pollution, which cause great concern to populations inhabiting shore areas.

ENVIRONMENTAL ISSUES AND SOCIETY

Emotionalism over environmental issues, even misinformed and distorted emotionalism exercised by some environmentalists, has made a real contribution to the improvement of our environment. The radical save-our-environment movement has shaken up legislators, government bureaucrats and corporation executives as could few rational scientific arguments. Congressmen listen to the Rachel Carsons, the Sierra Clubs, the loud environmentalists, because they are loud and persistent. But eventually a rational scientific basis must be considered if society is to solve its environmental problems in a lasting manner.

During the years from 1968 to 1973, concern over the pollution of our air and water, the depletion of our natural resources, and the protection and preservation of our environment grew from the province of a small group of scientists and private environmentalists into a dominant issue of our time. But problems generally included under the words "pollution," "ecology" or "the environment" are not new problems. As we shall see, since the earliest stages of its cosmic history the earth has been subject to ecological changes often of a radical nature unrelated to man's activities. All past civilizations have been dependent on, and concerned with, their environments. Civilizations that unwittingly ruined the fertility of their soil have vanished from the earth. "All the rivers run into the sea; yet the sea is not full; unto the place from whence the rivers come, thither they return again" is an ecological truth noted by the Preacher in Ecclesiastes. The farmer, the fisherman, the hunter were always environmentalists.

20

With the advent of the industrial age, man's use of the earth's natural resources has increased exponential ly and man releases over one million different waste products into the natural environment. However, few people were concerned with environmental problems until the late 1960's, when "ecology" became a popular word and the flood of environmental news and books began in earnest. There are literally thousands of environmental groups operating in the United States today. Now "ecology" is so common a word it cannot be avoided, even by the semi-literate. We have "ecology" products in our supermarkets, "ecology clubs," "ecology courses," and "ecology drives" in the community.

Why have the issues of environmental pollution and environmental protection come to the fore at this particular time? At best, there has been a kind of public awakening to awareness of the real dimensions of a real problem, since unbridled pollution could kill us all. At worst, this public concern, like so many others, will be short-lived. The pollution issue may salve the moral conscience of people yearning for something to distract them from other more difficult problems such as Vietnam, international political tensions, demands for equality, unemployment, the continuation of the nuclear-arms race, and global poverty. Perhaps the truth lies somewhere in the middle.

THE LIMITATIONS OF SCIENCE IN ENVIRONMENTAL ISSUES

One finds eminent scientists on both sides of almost any environmental issue. It is small wonder that the layman is often confused and it is inevitable that bureaucrats and officials have to make environmental judgments based on other than strictly scientific bases.

Distrust of science has heightened in recent years. In the eyes of much of the public, science has fallen from a position of preeminence among disciplines to a low estate: its practitioners are scorned as narrow and short-sighted; its benefits are described as destructive to human life and the environment.

In 1971, William D. Ruckelshaus, Administrator of the Environmental Protection Agency, was called on by some environmentalists to ban the use of 2, 4, 5-T, a herbicide that has been used for many years by farmers and gardeners. Ruckelshaus asked the National Academy of Science (NAS), the highest priesthood of U.S. science, to look into the matter. The NAS convened a prestigious group of scientists, investigated the matter and advised Ruckelshaus that 2, 4, 5-T should not be banned. In its reports, the advisory committee on 2, 4, 5-T concluded that "Current patterns of usage of 2, 4, 5-T and its known fate in various compartments of the environment, including the plant and animal food of man, are such that any accumulation that might constitute a hazard to human health is highly unlikely." After receiving the report from the NAS, Ruckelshaus banned the herbicide.

Scientific evidence and the advice of scientists are often ignored in environmental matters that have emotional or political implications. In the case of 2, 4, 5-T, it was charged that the herbicide caused birth defects, a highly emotional issue. From a bureaucratic point of view the simplest course is almost always negative; in this case, to ban 2, 4, 5-T. Whenever a new product or idea is introduced, the entrenched bureaucratic mind rejects it out of hand because it may not work smoothly, or it may rock the boat. This is especially true of many environmental issues where possibly long-range effects may occur at unknown levels of contamination, and there is no scientific basis for making a decision.

Many environmental issues are not susceptible to strict scientific evaluation since the long-range effects may be unknown. Low level contamination from DDT, mercury, airborne lead, noise and hydrocarbons all fall into this category. Many issues raised by science-based technology cannot be settled on purely scientific grounds: the issues must be either settled or ignored on economic, political and moral grounds.

POLLUTION IN PERSPECTIVE

Despite the ditherings of some of the children of Silent Spring, the inescapable fact is that the well-being of industrial man has improved as his environment has deteriorated. In perspective, the overall effect on human health even as industrial pollution has mushroomed in the 20th century, has been a marked increase in the life expectancy.

I am not arguing here that pollution is good for people, but rather that it is a relatively minor nuisance compared with other causes of death and unhappiness such as war, cigarette smoking, and alcoholism. No one I know is for pollution; if we had our way we would eliminate it. Still, it is clear to me that some forms of pollution are concomitant with modern life and industrial man must learn to accept some contamination of his environment, just as our "uncontaminated" forefathers living in a simpler era had to bear horse manure in their streets, unhealthy chilly winters, hot and sticky summers and lack of mobility.

The proposed ban on DDT is an environmental issue touching particularly on the question of its relative value to man and beast. Laying aside for the moment the name-calling, the charges of chemical-company profiteering and even the thin egg shells of those relatively few brown pelicans and their appealing feathered friends, it is wise to keep in mind the key fact relating to DDT and human health: despite widespread and extensive attempts to prove that DDT is toxic to man, it has never been shown that DDT when properly used causes human death or even illness. And in perspective DDT has increased the number of wild animals in areas where it has been properly used to combat insect pests. Barry Commoner would undoubtedly object if a farmer told him what chemicals he could or could not use in his St. Louis laboratory; and since Dr. Commoner does not claim to be an agricultural expert, it would seem wise and proper to allow the farmer to select his own methods for growing crops, methods that appear to favor the use of DDT.

I have high hopes for the environment of industrial man because steps are being taken to reduce pollution, and because nature has a way of taking care of herself. Consider a few examples of nature's ability to moderate short-term excesses: 1. Increased heat and airborne particles over industrial cities trigger increased rainfall, which tends to wash the air clean. 2. Increased use of large lakes and oceans as proper receptacles of biodegradable wastes results in increased growth of plankton and marine life. Lake Erie, often described as "dead" by ill-informed environmentalists, supports more fish than all the other Great Lakes combined, a fertility due in large part to the human and agricultural runoff into the lake. 3. Oil spills in the open ocean are converted by bacteria into food for marine organisms, and much of the oil dumped on that 95 percent of the ocean that is a virtual biological desert helps to sustain what life exists there. Numerous other examples of nature's ability to recycle wastes could be cited.

Manufacturers do not produce phosphate detergents in order to benefit the environment, but increased fertilization of infertile waters by phosphate nutrients has undoubtedly increased biological activity and fish in these waters. It is equally difficult to believe that manufacturers prefer phosphate detergents to soap simply because they can make a greater profit, as has been stated by many environmentalists. A large company geared to mass merchandising can sell almost any product, no matter how worthless or deleterious, at a good price to a lot of susceptible people; soap, detergents, cornflakes, cigarettes, lipstick, automobiles; the product per se is unimportant. Thus it is simplistic to claim that manufacturers are purposely pushing anti-environmental rather than pro-environmental products purely because the former are more profitable than the latter. Again one must closely examine the relative benefits and diseconomies of a product before calling for its elimination. Such examination often involves many unknown factors and requires careful analysis. And even when we are fairly certain that a product harms the environment, political, economic or religious factors may override our desire to eliminate it. The call by President Nixon for an increase in automobile production, when we know that the automobile is a prime desecrator of our environment points up the complexity of relative human values.

Many of us are not ready to accept logic regarding the relativity of environmental dangers because we have

24

been led to believe that our very existence depends on exorcising the demons of pollution, and no matter what, it is desirable to exist. Most environmentalists are not ready for ecological relativity. It is ironic that environmentalists belong to that segment of the population which lives well because the environment is being exploited to provide them with suburban homes and automobiles, two of the main causes of environmental deterioration in wealthy countries. We need an Einstein of ecology who can show us how to maximize the earth's natural resources in order to benefit the greatest number of people, and then we need to put his findings into practice.

CITIZEN INVOLVEMENT

Some ten years ago sober community leaders scoffed at "the birds and bees" people and went about their business of increasing the Gross National Product. This attitude has changed radically, now, and environmental issues have tremendous political clout. But citizen involvement in environmental questions has grown in the face of the lethargy of both business and government bureaucracy, and so have charges that environmentalists were against "progress" and more jobs.

The establishment refused to take environmentalists seriously and if it did not ridicule them, it managed to deny them a forum for their concerns. There was no way a citizen could legally involve himself in environmental cases since it was argued by the courts that the individual citizen or environmental group had no legal standing in pollution cases. Environmentalists in the early 1960's labored under the added difficulty that they did not fully understand how society makes decisions and carries them through.

Now the tide has turned in favor of the environmentalists. The American of the 1970's is increasingly concerned with personal spiritual fulfillment and quality of life. He has much more leisure time to study the broader issues of the environment and the meaning of life. The present furious public concern over environment will continue for some time to be an issue on which nearly everyone holds opinions, sometimes passionate opinions often based on deep-seated emotions rather than scientific fact.

Before 1965 U.S. citizens had no valid way to present environmental points of views and were forced to turn to crude, often loud demonstrations in order to be heard. Now that laws have been passed giving citizens and groups the right to take part in environmental legal proceedings the need for rambunctious demonstrations has diminished.

Speaking like a true hippie aesthete with few responsibilities and with no need to scratch for a living, Thoreau described the world of the well-to-do environmentalist:

This curious world which we inhabit is more wonderful than it is convenient, more beautiful than it is useful; it is more to be admired than used.

Our curious world is indeed beautiful, but modern society cannot afford the luxury of simply admiring it without using it. To retain its beauty we must learn to manage the world environment more wisely than we have since the time of Thoreau.

2

EVERYBODY AND NOBODY POLLUTES

As you know, the environment movement, right up to the first Earth Day, was a lovefeast with no enemies. After all, who could be for pollution? Now it believes, with Pogo, that we have met the enemy and he is us, meaning that there is an enemy after all, and it is the total population. I am really weary of seeing that Pogo quote on every piece of environmental literature. It is too easy to blame everybody, because if everyone is guilty, no one is guilty.
Jerome Kretchmer, July 5, 1971

NATURE: THE PRIME DESTROYER AND DESPOILER OF THE ENVIRONMENT
"We cannot command nature except by obeying her."
    Francis Bacon
    Throughout the long history of our planet, nature has wreaked incalculable destruction upon its air, waters, land and life. Arbitrary, capricious natural forces and phenomena, earthquakes, forest fires, volcanic eruptions, sea-level changes, erosion, climatic catastrophes, have brought about natural disasters of great and small magnitude. What is man in all this?
    The biggest natural cataclysmic change in the history of the planet Earth took place about two billion years ago when the whole Earthean atmosphere changed from a reducing to an oxidizing atmosphere. Due to the appearance in the evolutionary scheme of photosynthesizing green plants that liberated oxygen, a most toxic substance for anaerobic (oxygenless) life forms, most of the living things that existed at that time were eradicated. New types of life developed that could survive in an oxidative environment. Geochemical processes took on new characteristics, based on the slow oxidative degradation of both organic and inorganic materials. For most algae and bacteria

that had been living in an oxygen deficient atmosphere eons ago, the change to an oxidizing atmosphere was a form of lethal pollution.

Before Prometheus gave fire to man, carbon dioxide and carbon monoxide from enormous forest fires polluted the planetary atmosphere. It is surmised that lightning (hurled to earth by Zeus) often started the fires, and since Smokey the Bear was not around to put them out, the fires spread destruction and air pollutants over large continental areas.

During the International Indian Ocean Expedition in the early 1960's, British and Russian oceanographers chanced upon a patch of sea twice the size of Portugal, containing 40 to 50 million metric tons of dead fish, equal to the yearly world catch at that time. Since there is little human industry nearby, the fish could have died only from natural causes.

These examples of natural devastation remind us that man's efforts to pollute the planet's air and water are still puny compared to the miscalculations, if they were so, of nature herself. Time and natural change have extinguished more animal species than now exist on earth. We will not presume to ask why, only to note that nature over the millennia has destroyed more life, and polluted more environments than can be conceived of by most of her naturalist friends.

Man's simplest activities, his survival needs, such as breathing, eating and defecating, cause pollution. Perhaps, then, if we do not personally want to pollute the environment, we should eradicate man, but even in the absence of mankind the environment still would "pollute" itself, at least from the anthropomorphic point of view.

WATER

As a polluter of water, man cannot claim to compare with nature. Rivers were dirty (that is, full of sedimentary contaminants) long before man started adding his piddling pollutants. The mighty Mississippi carries a load of more than two million tons of sediment a day, sediments flushed from the land without much help of man, into the Gulf of

28

Mexico. Rivers are nature's transport system for sedimentary and organic wastes. Other shorter but less sluggish rivers carry a much greater quantity of "pollutants" to the sea. And when a river changes its course, it pollutes the estuary by bringing more sediment into it. Man has not appreciably added to this burden except in a few locations.

Man has little effect on the total quantity of rain that falls on the United States, carrying with it two-and-one-half million tons of sodium sulphate, an equal amount of calcium salts and carbonates, and thousands of tons of naturally occurring sulphuric and nitric acids. Many of these salts originated as nuclei for raindrops over the oceans, which were then carried by winds over our land.

So-called "natural springs" are often loaded with acids, salts and radioactive particles many times higher in emissions than the standards set by the U.S. Public Health Service. These high concentrations ensue when water leaches chemicals out of surrounding soils and rocks.

Animal life in a body of water depends directly on the amount of dissolved oxygen in the water. Oxygen in water occurs either as an effect of diffusion from the atmosphere or as a result of the photosynthetic activity of aquatic plants. If the level of organic material in the water is relatively low, the dissolved oxygen content is usually sufficient to oxidize the organics without causing oxygen depletion. However, if the organic load is too high or if productivity of the water causes an overabundance of plant cells (which during the night consume rather than produce oxygen), then either temporary or complete oxygen depletion of the water may take place. Since aquatic animals must have oxygen in the water to survive, even temporary oxygen depletion can result in death of animal life. Corpses of the aquatic animals sink to the bottom and add to the already existing organic load; then the short-term chances for the natural recovery of the water is small; but if oxygen again enters the water, animals may again appear.

Natural oil seeps release tons of crude oil on the land and in the oceans each year. For three centuries oil seeps have been observed off the California coast. Presently several natural seeps in the area unloose oil into the sea at the rate of about one hundred barrels per day, a rate greater than that of the Santa Barbara spill.

AIR

Nature is also the main polluter of our planet's air. Many plants emit hydrocarbons. The sweet smell of hay is caused by hydrocarbons. Trees, particularly conifers, emit terpenes into the air. Terpenes, isomeric oily hydrocarbons in concentrated quantities, which often appear in the air above a conifer forest as a haze, are poisonous to many forms of life. The Smoky Mountains are so named because of the natural terpene haze emitted by the conifers growing on their slopes.

Other hydrocarbons entering the atmosphere are released during the combustion of coal and oil by man, but also through natural production of methane, or marsh gas. Methane or "swamp gas" is produced naturally in tremendous quantities by the disintegration of biotic material in lakes, rivers and swamplands around the world. It is estimated that nature generates 1.6 billion tons of methane a year; man, less than 100 million tons.

The question of where the components of planetary air come from, how long they remain in the atmosphere and where they go to was investigated by Doctors Robert Robbins and Elmer Robinson of the Environmental Research Department of the Stanford Research Department, California. They found that each year about 220 million tons of sulphurous gases enter the world air mass in the form of hydrogen sulphide, sulphur oxides, and sulphuric acid. The prime sources of sulphur contaminants are volcanoes and natural emissions. Man discharges into the atmosphere only about one-third of the total amount.

Nitrogen compounds are the most plentiful of all "pollutants" in the air, and also a necessity for life on earth. Nitrogen gas normally comprises 80 percent of the atmosphere and nitrogen is continuously being converted to ammonia, oxides, or nitric acid by naturally occurring lightning, biological action and microbial synthesis. Man puts about 50 million tons of oxides of nitrogen into the atmosphere; Nature puts 500 million tons of nitrogen dioxide, 5.9 billion tons of ammonia, and one billion tons of nitrogen into the atmosphere. Natural lightning creates ten times as much airborne nitric acid as does man.

After studying the literature on sulphur in the environment, W. W. Kellogg and his associates reported in Science (1972) that man is now contributing "about one half as much as nature to the total atmospheric burden of sulphur compounds, but by 2000 A.D. he will be contributing about as much, and in the Northern Hemisphere alone he will be more than matching nature." The situation is aggravated in the vicinity of industrialized areas.

Changes in the oxygen levels in the air may result in the accumulation of toxic substances. Ultraviolet irradiation reaching the atmosphere produces ozone, which is an extremely active oxidant. The ozone molecule consists of three atoms of oxygen compared with the usual grouping of two atoms found in normal air. Ozone may react with sulphur dioxide to produce highly corrosive sulphur trioxide. This gas combines with atmospheric vapors to produce sulphuric acid, which is harmful to practically everything on the earth's surface.

SOME SHORT TERM NATURAL POLLUTION

During 1970, a volcano erupted in Iceland, spewing out poisonous fluorine gas that killed 7,500 sheep and lambs. Nature continually produces corrosive acids in large and deadly quantities. For example, tons of acid fumes emerged from Alaska's Valley of Ten Thousand Smokes, destroying plant life a thousand miles away.

Even simple rain can be considered an ecological pollutant if it falls in unusual ways and quantities. In 1969, rain fell earlier than usual in Victoria, Australia, triggering an abnormal cycle of abundant food for mice. The local mouse population exploded to "an estimated mouse density of more than 200 per acre." Farmers trapped as many as 300 to 400 mice in one night. The mice were said to be like a moving carpet in some areas, according to the Smithsonian Institute report on Short Lived Phenomena for 1970. Oysters and other marine organisms acclimated to the brackish waters of estuaries die when excessive rainfall dilutes the water of their habitat below minimum levels of salinity.

31

"Isn't there any other way for the gods to show their anger besides adding to the air pollution?"

Litchy
"Grin &
Bear It"
Washington Post

During the early summer of 1971, the "red tide" invaded the waters of South Florida. The red tide does not originate in the Kremlin; rather, it is a growth of masses of dinoflagellates, microscopic planktonic organisms that give sea water a reddish hue and exude a poison that is deadly to most fish and man. The Florida tide, caused by the organism Gymnodinium breve, comes and goes periodically for unknown reasons. When it appears, it often kills millions of fish. More than 500 tons of fish died during the 1971 St. Petersburg, Florida outbreak and over 50,000 tons were destroyed in the Gulf during the 11-month outbreak in 1946-47. Outbreaks of similar species of this natural pollutant have been recorded in India, Africa, California, the New York Bight, New England and scores of other places.

We all wish to be in harmony with nature; but it is apparent that nature is hardly benign, and not always in harmony with itself. Cataclysmic pollution from natural causes has marked the history of the planet Earth, and it continues today. When we study environmental pollution we must bear in mind that man pollutes the environment no matter what he does, but to a lesser extent than does nature. It is possible to minimize, but not to completely eliminate man-made pollution, and though we may diminish many of the deleterious effects of pollution, we can never return to a static, completely "natural" condition since there is no such ideal state.

"NATURAL" FOODS: NATURAL POISONS

During the past century chemists have cooked up a large variety of new compounds, some of which have found their way into our foods. Nitrates and nitrites are deliberately mixed into sausages so that the frankfurters and salamis will retain their nice red color; propionates are added to bread so that it does not appear spoiled though it stands on grocery shelves for weeks and then is foisted off on a customer as "fresh." SO_2 is put in soft drinks; the list is long and varied.

Risks from poisoning were for eons confined to workers in certain industries and victims of deliberate poisoners. Now whole populations use poisonous substances in the household, in industry, and in pharmaceuticals, and their use is growing. At least 30 substances commonly found in the house can cause death if improperly ingested (Bodin/Cheinisse).

A large group of other substances has entered the human food chain inadvertently. Some have been accused of various harmful effects: DDT of being carcinogenic (cancer producing); 2, 4, 5-T of being teratogenic (causing deformation of the embryo); and a host of others have been accused of simply being toxic, poisonous to the point of death. But since the life expectancy of industrial man has been rising steadily during the 20th century, he must be doing something right, including eating.

Though man has created new poisons, nature does not want for poisons, carcinogens and teratogens. Poisons of natural origin are numerous and were the cause of 90 percent of all poisoning in the 18th century. Many are still around and still killing people.

In times of social anxiety, peculiar food fads and freakish behavior often arise. In the 1960's America witnessed the rise of the Jesus freaks, the drug culture, a morbid interest in astrology, preoccupation with peculiar sexual perversions, and a great upsurge of the sale of so-called natural foods. Some of the earlier natural food fads, such as the macrobiotic yin and yang diet, which consists mainly of brown rice, have already peaked out, but the stress the environmental movement puts on chemical "poisons" in food has spurred the sale of so-called organically grown natural foods.

The caffeine in coffee and the tannin in tea have been shown to produce numerous cancerous tumors. An overdose of garlic will cause cystitis; bitter almonds and the kernels of apricots and cherries contain cyanogenetic glucosides, which cause cyanide poisoning. Spices such as nutmeg, parsley and dill contain highly toxic myristicin and apiole. Deaths have been caused by eating two or three nutmegs. Red pepper disturbs digestion and can cause diarrhea. Rhubarb stems are edible but the leaves, which contain high concentrations of oxalic acid have proven fatal. German raisins and vitamins A and D have caused birth defects in laboratory animals. Green tomatoes contain poisonous atropine. Honey not only has been shown to cause cancer when injected under the skin of laboratory animals, but will actually kill 50 percent of the animals which attain a dosage of one milligram per kilogram.

There is practically no substance, including salt and water, which will not kill if ingested in sufficient quantities. The problem in dealing with suspected or real poison is to find a realistic "no effect" level below which nothing happens, and a "frank effect" level above which poisoning takes place. The sad fact is that in 1973 we simply do not know what, if any, levels of pesticides such as DDT, or herbicides such as 2, 4, 5-T, are really harmful, if at all. It would appear that the present attempt to ban DDT is not for the protection of humans from poison, but rather for protecting approximately six species of predatory birds that in certain areas are endangered.

Organic food stores have sprung up around the country charging unconscionably high prices for "health foods" that are supposed to have been grown without chemical fertilizers, pesticides or herbicides. Purveyors of health foods claim that their products are nutritionally superior to commercial, chemically nurtured foods. However, most scientists who have looked into the matter in an impartial manner claim that there is absolutely no difference between organically and non-organically grown food. Health-food faddists, however, are convinced that their food not only tastes better, but also improves their health. This belief may in itself be a good reason for the continuation of their eating habits as long as they are willing to pay the high prices. Unscrupulous merchants, however, have been observed buying commercial produce from nearby supermarkets and then selling them as "organic" for twice the price to faddists.

The organic farming craze has brought about a peculiar distortion of values. Cow dung now costs more than cow milk on the Eastern Seaboard. According to a report in the Philadelphia Enquirer (February 14, 1972), "Milk is currently $6 per hundred weight (46 quarts)....Cow manure, packaged in five pound lots, brings nearly $14 per 100 pounds at garden marts."

1771207

POLLUTERS: THE HUMAN VARIETY

There is a tendency among many environmentalists to blame modern industry and technology for all of our pollution woes. In a sense they are right, for without modern technology the world would be half-starved, diseased, and lightly populated, as it was before the industrial revolution, and the quantity of air and water pollution much less than it is now. But that is like accusing a knife maker of murder when a felon stabs someone to death with his knife. Let us agree that we cannot go back to the "good old days" when the Thames and most other rivers near cities were open sewers and cholera, malaria and other "natural" diseases kept populations in check. Let us rather concern ourselves with the pollution problems we face in today's world.

We put the question: who and what has caused the addition of contaminants to our air, water, and land so as to pollute them? The answers are not clear-cut and may not please those looking for quick villains and easy solutions. It would simplify things if we could place the blame for pollution on industry only and declare industrialists the "enemy." But modern life is not simple and the sources and reasons for pollution in America today are varied and complex. Polluters are considered under the following categories: industry, power plants, commerce, agriculture, military, transportation, and the average citizen. In the latter category we include the single family house and the automobile, two personal objects beloved by many Americans.

INDUSTRY AND INDUSTRIAL WASTES

Industry is one of the greatest generators of pollution. Its waste contains all known air and water pollutants. The food, chemical, paper, and kindred bio-industrial groups generate, before treatment, about 90 percent of the BOD - Biochemical Oxygen Demand; a measure of oxygen-grabbing waste in water bodies - three times as much as is generated by the sanitary waste from the 210 million citizens of the United States.

Industry is by far the greatest cause of sulfur oxide and particulate pollutants. Most unwanted sulfur occurs as an impurity in fossil fuels. Sulfur dioxide can form corrosive sulfuric acid and descend to earth with rain water, resulting in an increase in the acidity of our lakes and rivers. Industrial particulates cause soiling but little damage to health. Both sulfur and particulates can be removed from industrial smoke stacks by modern technology.

Specific industries tend to emit specific types of air pollutants, as listed in the table below.

Integrated steel mills: Particles, smoke, carbon monoxide, fluorides.
Nonferrous smelters: Sulfur oxides, particles, various metals.
Petroleum refineries: Sulfur compounds, hydrocarbons, smoke, particles, odors.
Portland cement plants: Particles, sulfur compounds.
Sulfuric acid plants: Sulfur dioxide, sulfuric acid mist, sulfur trioxide.
Grey iron and steel foundries: Particles, smoke, odors.
Ferro-alloy plants: Particles.
Kraft pulp mills: Sulfur compounds, particles, odors.
Hydrochloric acid plants: Hydrochloric acid mist and gas.
Nitric acid plants: Nitrogen oxides.
Bulk storage of gasoline: Hydrocarbons.
Soap and detergent plants: Particles, odors.
Caustic and chlorine plants: Chlorine.
Calcium carbide manufacturing: Particles.
Phosphate fertilizer plants: Fluorides, particles, ammonia.
Lime plants: Particles.
Aluminum ore reduction plants: Fluorides, particles.
Phosphoric acid plants: Acid mist, fluorides.
Coal cleaning plants: Particles.

Despite grim predictions that the situation must grow worse because of increasing industrialization, the evidence indicates that air in most industrial cities is improving. The cleansing of industrial gas fumes does not present complicated technical problems; the main bottleneck to cleaner air is economic.

Industrial fluid wastes vary in quantity and composition from industry to industry and from one factory to another. The main forms of industrial wastes and their chemical types may be classified as organic, inorganic, and mixed, as indicated below.

INDUSTRIAL LIQUID WASTES	TYPE OF WASTE
ORIGIN	
Slaughter houses	
Fish and fish-meal factories	Organic compounds
Breweries	
Fat and margarine factories	
Fruit and beet sugar	
Dairy products	
Petrochemical	
Chemical pulp mills, paper mills	
Fiber and plastics industry	Mixed organic and inorganic compounds
Textile industry	
Leather industry	
Soap works	
Photographic chemistry	
Coal, iron, metal	
Mining	Inorganic compounds
Dye-stuff	
High explosives	

In the past, industrial fluid wastes have usually been treated by municipal waste treatment plants. Industrial plants that produce only a high BOD can safely discharge their waste effluents into local sewage systems which will only occasionally overload sewage plant facilities. New industrial plants producing high BOD levels often require the expansion of local treatment facilities. In the United States, Chicago and New York City levy surcharges on industrial establishments for treating their effluents. This reasonable practice is spreading to other localities and should induce greater compliance on the part of polluters.

Toxic industrial wastes such as non-degradable bio-chemicals, radioactive substances, acids, and concentrated heavy metals, cannot be channeled easily through a municipal

38

treatment facility. Such wastes either have to be treated on site, or special provisions made for their disposal. Large industrial complexes usually have their own waste treatment facilities whose effluents can safely be piped into municipal treatment plants. Thus, acids are neutralized with lime; metallic electrolytes are precipitated in the form of insolubles, and other toxic chemicals are reclaimed and recycled. Where no recycling facilities exist, special dumping sites are required for the disposal of toxic waste materials. Pollutants are either barged to sea or injected into natural or artificial underground cavities, such as abandoned mines. This latter practice, however promising, has created some unsolved technological problems; underground flows in the mine tunnels eventually exert such pressure on the dumped material that the closure-ceilings burst open, allowing the outgushing waters to wash out with the disposed-of waste. Such outbreaks will then produce acute environmental pollution, leading occasionally to the complete denudation of areas covered by the escaping water flow.

THE MINING INDUSTRY

Two specific industries, mining and petroleum, deserve special notice as polluters. Mining wastes generate particularly damaging pollutants in the form of high concentrations of both highly soluble inorganic matter and sediments, debris, and other fine suspended particles such as colloids. Acid mine drainage from coal mines pollutes more than 10,000 miles of streams in the U.S. Up to 70 percent of the drainage comes from abandoned mines, most of it from exhausted underground mines. Acid mine water is formed by the oxidation of iron sulfide or pyrite in a series of reactions involving sulfates, sulfuric acid and iron oxides. Slurries, solid particles suspended in water, are frequently used to transport ores through pipe lines where they are mixed with large quantities of water. Usually less than half of the slurry water is recovered for reuse; the rest is released into either settling ponds or streams.

Mine slurry waters introduce both dissolved and suspended matter into local water streams. Many heavy metal ions are highly soluble in water and arsenics, lead, and even silver compounds are frequently carried into the rivers. Concentrated brines of common dissolved salts are chemically and physically toxic to many aquatic organisms. Radioactive radium 226, with a long half-life, is a waste product of uranium mining operations, while arsenics are released from silver mining operations. Cyanide is used in the extraction processes of gold. The purification of table salt generates brines high in potassium chlorides. The mining industry attempts to remove highly toxic compounds in forms of "relatively insoluble precipitates." Radium is converted into radium sulphate, and arsenic into its carbonate form. Slight solubility, however, still can introduce significant quantities of toxic materials into the aquatic environment.

Mountains of mine tailings which, in most cases, contain radioactive and toxic compounds and other contaminants, are exposed to the elements. Rain percolates through the tailings, invariably dissolving and carrying pollutants into lakes and rivers. Treatment of most tailing overflows seems at this time to be both impractical and uneconomical. However, at the rate our natural resources are being depleted, and the price of raw materials is increasing, we should soon reach a point where tailings can be mined economically. The suspended solids may be slag, sluice sands, or clays washed from around the desired ores. They can cause turbidity in the receiving streams and may occasionally accumulate in river beds, slowing water movement. Due to a decrease in light penetration in highly turbid waters, oxygen production is simultaneously decreased. This reduces the regenerative ability of the waters and allows anaerobic bottom processes to develop, leading eventually to the release into the water of further noxious or outright toxic materials, such as hydrogen sulphide. Clay minerals may have a beneficial effect in the streams, since they serve as natural absorbents or flocculents for many organic and some inorganic compounds, and thus may improve the quality of the water by removing some toxic substances and nutrients from it.

THE PETROLEUM INDUSTRY

Because of the toxicity of the products it handles and sells, the petroleum industry is a prime polluter of the environment. Drilling for oil or gas results in suspended solids, which are washed away from the drilling site and released into the environment. Brine deposits are frequently associated with oil deposits and the highly concentrated salt solutions are often forced from the drill pipe by the released well pressure. When this brine enters fresh water it can cause osmotic stress and often death to the organisms present.

Accidental oil spills or uncontrolled leaks of oil and tar from refineries are harmful to some aquatic life, especially birds which frequent the water's surface. The oil may deprive water vegetation of adequate illumination and thus stunt its growth. Oil residues ingested by commercially valuable fish or shellfish, such as oysters, impart an oily taste which may render the products unmarketable. Varied refinery and petrochemical plant effluents cause a number of undesirable environmental effects: they deplete the dissolved oxygen content of natural waterways; contaminants such as phenols, arsenic, lead and cyanide are toxic to wild (and not so wild) life; they may impart undesirable tastes and odors to natural waterways. In addition petroleum product spills may be a fire hazard and almost always impair the aesthetic value of wilderness, waterways and beaches.

POLLUTION FROM POWER PLANTS

The growing use of electricity and the need for heat in winter has greatly increased the use of fossil fuels for generating electrical and heat energy. Coal oil and gas, the fossil fuels, are chemicals formed by a biological process: transformation of solar radiation by plants into carbohydrates that are slowly changed into combustible hydrocarbons by geological processes. Burning transforms the chemical-compound hydrocarbons by combining them with oxygen into heat energy, giving off carbon dioxide, water, and often trace amounts of sulphur, mercury, and carbon monoxide in the process.

World combustion of fossil fuels is likely to grow at a rate of 4 percent per year, which would cause a doubling of demand for fossil fuels every 17.5 years. Despite a shift towards nuclear power in the developed nations, fossil fuel use will grow throughout the world. Billions of Asians, Africans and Latin Americans will want, and get, automobiles and electricity. Even if population were not to increase, demand for fossil fuels will grow, and since it is becoming less efficient to mine coal and dig oil, more energy will be required to extract, treat and transport fossil fuels.

In 1971 the world burned fossil fuels to generate 5,900 billion watts of power (Frisken, 1971). Our planet receives 176,000,000 billion watts in the form of radiant energy from the sun, of which 50 percent, or 15,000 times as much as is generated by man, reaches the Earth's surface. One hundred twenty years from now we will be generating 1/7,500th of the energy received from the sun, an amount that is not likely to cause climatic changes. Actually, no one knows for sure when, if ever, man's dissipation of thermal energy into the environment will cause serious atmospheric damage. So far, man's industrial activities have not changed the global climate, although they have affected temperatures and rainfall in the vicinity of some industrial centers.

Though other forms of energy appear to have environmental appeal, fossil fuels are still more economically attractive and are likely to be used until they run out. In 1971 the world's proven reserve of oil was 500 billion barrels, of which one-fifth is offshore. Eventually, probably in less than 100 years, the wells will run dry. Gas and coal deposits will probably last two or three centuries longer, but after that man will be forced to turn to other forms of energy.

To put power plant emissions in perspective it is important to realize that they represent only about one-seventh of all air pollutants. In 1970 the United States used and converted into waste heat 630,000 million million BTU's (British Thermal Units) of energy. A BTU is the amount of heat required to raise a pound of water one degree fahrenheit. Of this enormous quantity of energy

42

34 percent was used in homes, primarily for space heating. About 20 percent was consumed by cars, trucks, buses, trains and planes in the form of liquid petroleum products. Power plants used 11 percent; other industry used about 24 percent of the total; 10 percent was used by commercial establishments, and only about 1 percent was used by farmers for tractors and irrigation purposes. Over 90 percent of all energy came from fossil fuels. Hydro-electric power plants now generate about 20 percent of all power. Nuclear power generation is still relatively insignificant, but is likely to grow rapidly in the next 20 years.

Five major air pollutants result from the burning of fossil fuels: carbon monoxide, sulfur oxide, hydrocarbons, oxides of nitrogen and suspended solid particles.

Lead added to gasoline as an anti-knock agent enters the air when the gasoline is burned. Mercury and other heavy metals are spewed into the air from certain fuels, particularly coal. Asbestos fibers, which are highly carcinogenic, float into the air from construction projects and automobile brake linings.

Almost all fossil and nuclear power plants cool their condensers with water, and in the process reject large quantities of waste heat into the cooling waters. The temperature and quantity of rejected heat depends on the sizes and efficiency of the particular plant. Waste water from power plants is usually about 15° F. higher than when it entered the condenser. These discharges, when entering a body of water, will warm the water in the vicinity of the discharge pipe and may change or kill aquatic biota.

COMMERCE

The buying and selling of goods in commerce generates an enormous amount of packaging wastes but relatively little in the way of air or water pollution. Packaging accounts for 40 percent of the non-construction solid waste load in New York City and many other industrial communities.

Most commercial wastes originate mainly from the way products and materials are packaged. Non-returnable packaging has decreased the tendency on the part of society to re-use packaging materials. If goods are not recycled,

Table 2-1, taken from the Council on Environmental Quality's 1972 report, indicates that power plants are the leading source of sulphur oxides being added by man to the air over America. However, since many coal burning power plants are in isolated parts of the country, much of their pollutants are not breathed by many humans, though the oxides may increase the acidity of lakes and streams. Industrial sources—mainly pulp and paper mills, iron and steel mills, refineries, smelters and chemical processing plants—add more particulates to the air than any other single source. The 100 or so million motor vehicles on American roads contribute more carbon monoxide, hydrocarbons and nitrogen oxides than all of the other sources combined. However, it is noteworthy that spaceheating of homes and other buildings generates pollutants in close proximity to large concentrations of people and hence may be a more serious aesthetic and possible health problem than other apparently larger sources whose pollutants are more dispersed.

Table 2-1   Estimated Emissions of Air Pollutants by Weight, Nationwide, 1970 (Preliminary Data)

in millions of metric tons.

Source	CO	Particulates	SO_2	HC	NO_2	TOTAL
Transportation	111.0	0.7	1.0	19.5	11.7	143.9
Fuel combustion in stationary sources	.8	6.8	26.5	.6	10.0	44.7
Industrial processes	11.4	13.1	6.0	5.5	.2	36.2
Solid waste disposal	7.2	1.4	.1	2.0	.4	11.1
Miscellaneous	16.8	3.4	.3	7.1	.4	28.0
Total	147.2	25.4	33.9	34.7	22.7	263.9
Percent change 1969–70	− 4.5	−7.4	0	0	+4.5	

Source: Environmental Protection Agency.

44

either because of non-reusable packing materials or because the products must be replaced after a very limited lifetime, production is sustained at an artificially inflated high level. Thus, the trend to relatively cheap throw-away materials for packaging, such as paper, aluminum or plastic rather than reusable containers.

Artificially reducing the useful lifetime of items or products creates massive quantities of wastes in the form of scattered junk and discarded products. Advertising material in the form of masses of paper trash amounts to approximately one hundred pounds per person per year. City sanitation departments are finding it difficult to cope with the spreading of leaflets, give-away items, and advertising mailers, all of which add to the high levels of garbage.

Who is to blame for products with short lifetimes, the public which buys and uses them, or the commercial interests which sell them? Except for sustaining high levels of production, there is little reason why the usable life of an automobile should be only four or five years instead of twenty, and the same holds true for practically all modern household appliances. Over sixty million discarded automobiles are piled up along U.S. highways, streets, parks and riverbanks. Junk cars increase by two and a half million each year. Most cities are unable to cope with this problem, although automobile shredding machines have recently been developed, and a few are in use. Compacting of junked vehicles for re-use in the manufacture of steel is economically feasible only on a very large scale. In order to re-use the vehicles, their interiors usually must be burnt out before they are compacted. Since plastics are extensively utilized in automobile manufacture, and the burning of plastic produces noxious gases laden with fluorides for which no efficient air pollution abatement methods have yet been devised, junked vehicles can not easily be disposed of without causing some air pollution.

AGRICULTURE AND AGRICULTURAL WASTES

Large-scale agricultural activities cause a number of serious pollution problems. Solid wastes from the 49 million cattle, 57 million hogs, 3 billion poultry and 21 million sheep in the U.S. amount to about 1.3 billion tons annually, equivalent to the waste of 1.3 billion suburban Americans. In addition, animals generate 700 million tons of liquid waste.

Agricultural waste problems are particularly severe when the animals are kept in confined areas, as they are in modern feed lots and chicken factories. Despite the fantasies of organic farmers, these wastes have limited agricultural value compared to chemical fertilizers. Piles of unwanted manure create serious soil, water pollution and odor problems as well as provide breeding grounds for insects.

Food-processing plants create unique waste problems. Effluents from canneries, slaughter houses, dairies and refineries are particularly high in BOD, suspended solids, nitrates and phosphates.

Runoff from crop lands carries large amounts of fertilizers and pesticides into receiving waters. In the United States fertilizer use during 1969 was approximately 7 million tons of nitrates and 5 million tons of phosphates. Widespread pesticide use on farms and in suburban towns results in some leaching of long-lived pesticides into nearby waters. In addition, one to two billion tons of sediment from crop lands flow into U.S. streams and lakes each year. Sediment particles often clog slow-moving rivers and accelerate eutrophication of lakes.

MILITARY

Simply from an input-output point of view, the military forces in this and other countries have produced, if that is the proper word, an enormous quantity of lethal toys which litter the landscape. Not all of the $80 billion spent each year by the Pentagon goes for the upkeep of generals, admirals and their retinues. Only about $25 billion a year has lately been "disposed of" in a truly military manner in Southeast Asia. What remains is a vast array of military hardware, poison gases, atomic hardware, lethal biologicals and other weapons that constitute environmental hazards and a blight on the landscape.

During the 1945-1963 period, military testing of nuclear weapons caused a profound jump in radiation levels in the atmosphere and the biosphere. Radiation levels were found in milk and many other foods increased as radio-active particles came to earth with the rain. For the moment, the U.S., Great Britain, and the U.S.S.R. have discontinued above-ground testing of nuclear weapons, but France and China continue to test, though at a much lower level than the first "dirty" experiments.

Below-ground testing of weapons may also cause damage to the environment and may help trigger earthquakes. In November 1971, a number of vociferous groups tried to prevent an AEC-military underground nuclear blast in Amchitka because of possible damage to the surrounding area. The blast did not create the damage predicted by the environmentalists. However, their argument that the blast might precipitate an earthquake was valid.

Over ten thousand discarded military airplanes from World War II sit on the ground in the Arizona desert, where the dry air keeps them from disintegrating. Old military explosives and chemicals are waste products that are uniquely difficult to dispose of. What does society do with billions of megatons of nuclear warheads and bio-logical agents that could destroy the human race? Military wastes are an environmental problem that has not been given the attention it deserves because of the aura of secrecy that covers many military operations and the unwillingness of the average citizen to tangle with the Pentagon.

The environmental effects of war have been demon-strated by the U.S. armed forces whose bombers and artillery have blasted over 25 million craters in South Vietnam from 1965 through 1971. This gruesome en-vironmental pocking is distributed over a land area the size of the state of Missouri. The Pentagon distributed some 26 billion pounds of explosives in Indochina, over twice the amount used by the U.S. during all of World War II. Many of the bomb craters have made farmland useless and formed stagnant waterholes where mosquitoes breed. A study by Professor A. H. Westing and E. W. Pfeiffer (1972) on the bombings concludes that there will

47

be long-lasting deleterious effects on the ecology of South Vietnam in the form of wasted farmland, eroded land, useless forests, increased mosquitoes, and contamination of water supplies. These effects have been brought about by explosives equivalent in destructive power to over 360 Hiroshima-type atomic bombs.

TRANSPORTATION
   Pollution resulting from the movement of goods and people is intimately bound up with all the other causes of pollution. Air pollution would be greatly reduced and abandoned cars would not litter the landscape if cars and trucks did not exist. Transportation is the lifeblood of commerce, and the military would be moving on beans and hay if it were not for modern transportation. Twenty percent of the gross national product goes for transport of goods and material ($200 billion). But the biggest cause of transportation pollution is the 110 million privately owned automobiles run by average American citizens. Among the unwanted offspring resulting from the American's love affair with the automobile are noise, air and water pollution, as well as solid waste.
   How does America move about? Of the money Americans spend on transportation 83 percent goes for cars and trucks. In 1970, Americans spent $94 billion on cars and only $16 billion on all other forms of passenger transportation: planes, buses and trains.
   Transportation of freight by trucks constitutes 74 percent of all freight expenditures in the U.S. In 1970 $64 billion was spent by shippers on truck transport compared to some $10 billion on rail transport and less than $9 billion on all other forms of materials movement. The railroads, however, moved more ton-miles of intercity freight than did trucks.

FREIGHT	METHOD	COST
Passengers	Cars	$ 94
	Planes, trains, buses	16
Commodities	Trucks	64
	Trains	10
	Pipelines, etc.	9
	TOTAL -	$198 B

48

In addition to the $94 billion spent on automotive travel, billions went to building roads, which have deleterious effects on the environment far beyond their main function of encouraging automotive travel. Over $243 billion has been spent on the federal interstate highway system alone, with at least another $30 billion necessary if we are to complete the remaining 10,000 miles of the 42,500-mile interstate network by 1977. Environmentalists have organized in many parts of the country to fight highway construction and a national group, the Highway Action Coalition, is fighting to eliminate the Federal Highway Trust Fund, the main source of funds for federal highway construction.

Transportation pollutants include most of the carbon monoxide and nitrogen oxides in our air as well as asbestos from the brake linings of cars and trucks. Asbestos particles when breathed are insidious pollutants: cigarette smokers exposed to high levels of asbestos particles have a 90-times greater chance of contracting lung cancer than does the average non-smoking citizen. Non-smokers who continually breathe asbestos particles tend to develop lung cancers 15 times as frequently as those who are not exposed.

Vehicles cause much of our city noise, an environmental stress that affects the nervous and endocrine systems. Noise causes blood vessels to constrict and may in some cases aggravate existing deseases. Even if it does not kill, unnecessary noise can cause aggressiveness, headaches, and loss of appetite and sleep. Airplanes emit hydrocarbons that can often be seen as an oily glaze on waters surrounding airports, such as the sheen on the waters of Jamaica Bay near New York City's John F. Kennedy airport. Supersonic transports (SST) could have disastrous effects on the upper atmosphere and on the eardrums of humans and animals within earshot of the supersonic boom. All these environmental delights are consequences of transportation.

THE AUTOMOBILE

Car ownership will probably increase for a number of years because of suburban sprawl and lack of adequate public transportation. However, it has become apparent that the car is a prime destroyer of our environment. Consequently, Philadelphia, San Francisco, Ontario and Boston, among other areas, have taken steps to install adequate mass public transportation systems and thus reduce dependence on cars. Governor Sargent of Massachusetts, once a staunch supporter of highway construction, remarked that "The question is not where an expressway should be built but whether an expressway should be built."

Without question, most air, water, land and noise pollution springs from our use of internal combustion motor vehicles. In the U.S. there is approximately one automobile for every two individuals. One result of this is pollution: 75 to 95 percent of air pollutants (by weight) in our major cities; essentially all of the lead in our air, and most of the lead in our waterways, lakes and oceans; most of the asbestos in city air; oil pollution in our oceans and waterways, which comes largely from waste lubricants and airborne petroleum products (much of the oil spilled during ocean transport is due to the insatiable demands of cars for gasoline); land pollution in the form of roads and garages for cars (one-third of Los Angeles is paved over to accomodate cars; in the U.S. an area equal to all the New England States south of Maine, plus Delaware, is paved over for automobile use); noise from horns honking, brakes screeching, and wheels grinding, blasting the air of most of our cities.

In 1972, automobiles spewed some 145 billion pounds of pollutants into the American atmosphere. Current federal legislation to limit pollution emission applies only to new car production and should, if Detroit and the oil companies cooperate, reduce present levels of automotive pollution to 130 billion pounds by 1980. Thereafter pollution levels will resume their rise because of anticipated increased use of automobiles, assuming that Americans will, by the 1980's, still not have kicked the automobile habit.

Cigarette smoking, another American habit, has been labeled a health hazard. According to the American Cancer Society, 21 million Americans have given up cigarette smoking, but 44 million Americans still smoke, and cigarette sales went up in 1971 and 1972. There is

MISONEISM TOPPLES U.S. DEMOCRACY

As misoneists took over the ecology movement, the stagnation caused by well organized
environmental groups became manifest.

Agricultural output decreased dramatically
in 1975. Farmers denied recourse to chemical
biocides lost their crops to insects and weeds.
Misoneical environmentalists maintained their
campaign against DDT on the grounds that it
caused the thinning of eggshells of a few predatory birds. Farm production declined despite
an increase in acreage which cut into recreational preserves. Food prices rose by 20% in
1977. Parklands bordering on swamps were a-
bandoned to growing swarms of mosquitoes,
flies, gnats, ticks and chiggers. Fire Island
and Walt Disney World were rendered uninhab-
itable.

Power plant construction came to a virtual
standstill and the Alaskan pipeline remained
unbuilt due to dilatory legal tactics of extreme
environmentalists. By 1974, frequent fuel and
power shortages were causing massive layoffs.
Production costs increased. The U.S. standard
of living plummeted as American technology
lost its competitive edge in international markets, and science research fell prey to misoneical budget cuts.

Growing discontent gave way to rioting by
consumers, workers and students. The President called on the Army to put down riots in
Detroit and Chicago. Unopposed by a fearful
Congress, the President declared martial law
and assumed extraordinary powers - in the
interest of national security. Thus did the U.S.
drift into dictatorship.

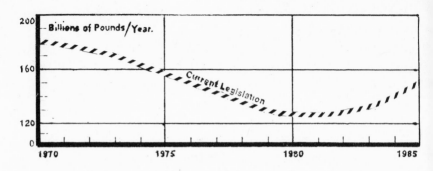

Figure 2-2 Air Pollution in the U.S. from Automobiles
Should Rise After 1980

Source: U.S. Environmental Protection Agency

reason to assume that a well organized campaign against
the automobile habit will produce equally disheartening
results.

In many ways the automobile habit is similar to the
cigarette habit. Both are extremely hard to break; both
may be based on irrational psychological needs. However,
cigarettes are an individual vice, whereas automobiles
are of necessity fostered by community action in the
form of road building and unnecessary dispersal of facilities. Both the cigarette and the automobile habit are
promoted by advertising and sold as contributing to "the
good life." Both cigarettes and cars are superfluous to
a healthy style of living, although one would never know
it from reading the advertisements in American magazines.

Once the individual has become "hooked" on cars or
butts, it is often difficult to break the habit. To many the
car represents sexual manhood and machismo. For youth,
it provides a place where wooing can go on free from
parental restraints. To some, the automobile gives the
illusion of mobility and independence. The poor (black,
red and white) see in the car a means of temporary
escape from a slum when their situation becomes in-
tolerable. A black person cannot always buy a house,
but he can usually buy a car. It is small wonder that
many blacks prefer to purchase cars before investing
in, or trying to buy a house.

The movement of families from cities to the suburbs has forced many Americans to own cars simply to move around. Suburbs require cars and cars encourage suburban life, a vicious cycle.

THE SINGLE-FAMILY HOUSE

Much of the deterioration of the American environment can be traced directly to the suburban single-family house. This phenomenon erupted spectacularly in the post-war world due to the opening up of the suburbs to white, middle-class home owners, a movement spurred by government subsidies in the form of FHA loans and tax abatements to home owners. The ready availability of automobiles made suburban living possible; it is more than coincidental that the great increase of pollution since 1950 parallels the mass buildup of sprawled suburbs.

The suburbs gobble up more space than other forms of housing since the average suburban family demands about a one-quarter lot. Therefore 100 homes require 25 acres plus roadways to each house. Those same hundred families living in one 25-story apartment house (four families per floor) would take up only one acre, and the remaining 24 acres could be used for parks, playgrounds and farms. In my neighborhood, the upper West Side of Manhattan, we have a population density of about 300-400 dwelling units per acre. Incidentally, I do not feel crowded, and the many ground-level stores and people on the street at all hours gives one a sense of security that is often absent in sprawled communities. Very often land taken by suburban single-family houses is usually the most valuable open space because it is close to large cities.

The single-family house uses heat and materials less efficiently than an equivalent apartment dwelling. In the suburbs most houses require the use of automobiles, which add considerably to the many pollution problems. Because it has a larger surface-area to living-space ratio than an apartment, the single-family house requires more energy to heat and cool than an equivalent apartment. This leads to greater waste of energy and consequently more air and water pollution.

CITY AND SUBURBAN WASTES

Let us compare the pollution and waste generated by an average family in Manhattan (New York City) and a similar family in nearby Nassau County. Both communities have a resident population of about 1.5 million persons. Manhattan and Nassau are only 30 or so miles apart, but the life styles in the two communities are quite dissimilar. The average Manhattan family lives in an apartment house and does not own a car; there is an average of one car for every four households, or about one registered vehicle for every eight inhabitants. The average Nassau family lives in a single-family house on a quarter-acre lot, with a lawn in front and a yard in the rear, and owns and operates 1.7 cars. New York County (Manhattan) is the smallest county in the U.S. Much of its 23 square miles is devoted to industrial and office buildings. Nassauians live on approximately 300 square miles, 17 percent of which is taken up by roads and garages; there is relatively little industry in Nassau.

Most American single-family houses require a considerable amount of wood for their construction. Apartment houses on the other hand use wood only for doors and cabinets and occasionally for flooring in the fancier buildings. Because the single-family house uses two to three times as much wood as does an equivalent highrise dwelling, two to three times as many trees must be cut down to build each single-dwelling unit. Forests must be destroyed to build the millions of new suburban dwelling units now in the planning stage.

SOLID WASTE

Assume both families eat the same quantity of food and read the same weight of papers and magazines. Still, the suburban family generates 10 percent more solid waste simply in the form of lawn clippings and garden debris. Also there is more junk around single-family houses in the form of rubber tires, old lawn equipment, woodworkings, old cars, and so forth, than around apartments. The 700,000 cars owned by Nassau families compared to the 200,000 cars owned by Manhattan families form a formidable solid waste load as they wear out.

54

Suburban living is inherently wasteful of natural resources. Consider the waste of steel due to excessive dependance by suburbanites on the automobile. The approximately 700,000 automobiles in Nassau, weighing 3,000 pounds each, require about two billion pounds of steel; an enormous rusting mass which will be solid waste in a few years. Additional solid waste in Nassau comes from use of garden equipment, home heaters, snow removal equipment and other paraphernalia associated with single-family dwellings.

NOXIOUS CHEMICALS
The Nassau family uses insecticides and herbicides on its lawns and gardens and in the process causes run-off into streams and ground water. Manhattanites garden mainly in window boxes. Nassau families with pools cause additional chemical pollution through use of algicides.

THERMAL WASTE
The average apartment requires less energy per family unit each winter than does a single-family house. Because the house has thinner walls and greater surface area per interior room than the average apartment house, it is much less thermally efficient and therefore requires considerably more energy to heat in winter and to cool in summer. The heat is usually supplied by an individual home burner which gives off air pollutants. Apartment house heaters are of necessity larger and usually more efficiently maintained than are home burners.

Compared with single-family houses or even individual apartment houses, central air conditioning and heating plants used in large consolidated housing projects greatly reduce thermal waste and air pollution. For example, Co-op City, a 35-building complex of apartment houses built in the Bronx during the late 60's, houses approximately 55,000 persons on 300 acres, most of which is landscaped. One central power plant generates heat during the cool months and refrigeration during the hot months. Using one plant instead of 35 required 20 percent less cooling capacity and reduced the initial cost considerably. In addition, one plant takes up less space than

35 smaller ones would have and is much less expensive to operate. Such economies and environmental advantages of scale are possible with well-planned high-rise dwellings but not with single-family homes.

AIR POLLUTANTS

Non-industrial air pollutants come primarily from burning fuel and from automobiles. Since the single-family house requires more fossil fuel to heat than an equivalent apartment, it generates more air pollution directly.

In addition, the Nassau County family generates more than three times as much air pollution as does a city family because of the extensive use of cars in the suburbs. Most Manhattan housewives can walk to local grocery and service stores; housewives in Nassau hardly ever walk because of the great distances between homes, shopping centers and schools. Manhattan kids usually walk to public school. Nassau kids are bussed or driven, again because of the general sprawl.

The average Nassau family travels more than 20,000 miles per year by car. Assuming they own a car that meets the 1971 standards for emissions (hydrocarbons: 2.2 grams per mile; carbon monoxide: 23 grams per mile), they will generate 44,000 grams of hydrocarbons and 460,000 grams of carbon monoxide during 1971. Typical vehicles with uncontrolled emissions give off more than three times these "acceptable" values.

Despite the fact that suburbanites generate more air pollutants than do city dwellers, concentrations of most air pollutants are usually greater in the city. The relatively high levels of air pollution on some Manhattan streets are due mainly to the seven million vehicle miles per day traveled by commercial traffic, taxis and out-of-town automobiles, and by the greater concentration of pollution sources. Trucks contribute from one-half to three-fourths of the air pollutants in Midtown Manhattan streets and taxis emit more than one-third of Midtown pollution.

WATER POLLUTION

Because of lawn and garden watering, individual home swimming pools and car washings, the average Nassau family uses and wastes much more water than the average Manhattan family. Only 50 percent of Nassau is sewered; 50 percent of households still use septic tanks. Wastes in these tanks eventually drain through the sandy soil towards the underground local water supply. Like most older established cities, all of Manhattan is sewered and most of its waste water is treated in sewage treatment plants.

In summary, compared to the average Manhattan apartmented family, the average suburban Nassau family generates:

—— more than three times as much air pollution

—— about 15 percent more solid waste

—— considerably more insecticide and pesticide run-off

—— at least 10 percent more thermal waste from home heating

—— much more waste water in the form of garden run-off

—— land pollution in the form of needless scarification of the countryside

—— greater waste of wood and other natural resources

An inescapable conclusion from this analysis is that this nation can materially reduce pollution by bulldozing Nassau and similar developments into parks, gardens and small farms, and by allowing the individuals now living in these split-level pollution factories to move into rational high-rise apartment communities serviced by mass transit facilities.

Who are the polluters? Everyone who drives a car, every power plant, every industry, every litterbug, every man, woman and child who breathes. But some of the pollution sources - especially industrial plants - are much more obvious targets and easier to control than others. Without a change in life styles it will be almost impossible to decrease the pollutants associated with individual automobiles and single-family dwellings. It will be relatively easy, though costly, to reduce the pollution from industrial sources. The technology and laws now exist for a concerted and successful attack on industrial polluters. In the process many smaller, less efficient plants will fall by the wayside and the cost of some goods will increase. Pressure from the thousands of anti-pollution groups and from concerned citizens throughout the country will inevitably result in a cleaner environment. But in a sense Pogo was right; the average citizen, by his choice of life style, has contributed to the overall pollution problem. Only a reordering of priorities will reduce many insidious forms of pollution.

An 1871 view of Wall Street by Thomas Nast

3

POLLUTION WARS AND THE SOCIOECONOMICS OF
ECOLOGY

"When the pollution-oriented health administrators and
the public alike begin to focus clearly on the enormity of
the bill that would be required to reduce pollution to meet
unnecessarily severe standards...precipitously prepared
from undigested, dubiously related facts...on which the
public has been ill-advised or misled..., then will come
the day of reckoning and rude awakening to the folly of
past anti-pollution actions."
H. E Stokinger (Science, November 12, 1971)

The hot and furious war against pollution has covered
the battlefield with armies of instant environmentalists,
industrialists whose initial reaction to any change is
negative, politicians eager to ride the bandwagon of the
victorious troops, and, of course, lawyers--the harbingers
of dissension. Both sides are often loud and vociferous,
each accusing the other of deceit, avarice and stupidity.
The pollution wars are likely to leave lasting socioeconomic scars on the countryside.
In this chapter we discuss environmental groups, some
of the actions they have taken and laws promulgated by
society to reduce pollution. An important component of
the pollution wars are the environmental impact statements and hearings required of all potential polluters
A vital environmental consideration is the cost of pollution
and of pollution control, an issue of great concern to
municipalities and firms asked to stop polluting, but an
issue often ignored by environmentalists demanding action.
Finally we give some examples of eco-pornography, the
often misleading advertisements used by some industrial
firms to show that they are "fighting" pollution.

ENVIRONMENTAL GROUPS

There are 750 environmental groups in New York State according to John L. Loeb, Jr.'s New York State Council of Environmental Advisors. No one really knows, but there are probably 5,000 to 10,000 non-governmental environmental groups in the nation as a whole; it is almost impossible to keep track of them all. Some ad-hoc groups form to fight local issues such as air pollution, a proposed power line or a construction project. A few groups are national in scope, well organized and well funded. The total membership of all environmental groups is probably between 500,000 to 1,000,000 -- excluding the National Rifle Association, which alone has over 1,000,000 members, and the 600,000 Camp Fire Girls. The resourceful Ralph Nader is seeking to unite 60,000,000 anglers into a group to fight against water pollution. It is hard to draw a line between groups that are essentially environmentalist and those that wish to live in a better environment but have other interests in life.

The well known groups have widely diverse memberships, and work in surprisingly different ways. To understand what is happening in the environmental field in America today it is necessary to know something about them. We have chosen for consideration a few representative groups: Nature Conservancy; Scientists' Institute for Public Information (SIPI); Sierra Club; Environmental Defense Fund (EDF); Council on Economic Priorities.

THE NATURE CONSERVANCY

During the past few years, the Nature Conservancy has probably done more to preserve undeveloped land than any other group in the country. This group does not engage in legal or publicity battles with "evil forces" -- it simply acquires environmental and ecologically significant land and preserves it in a natural state.

Ecologists now realize that in order to "preserve" endangered species one must preserve their habitats. It is not enough to keep a few representative samples in

60

zoos. In order to preserve wilderness, open land and shore property in the United States, it is necessary to keep the land out of the clutches of land developers. This can only be done by ownership of the land, either by a government body dedicated to preserving it, by wealthy individuals who have no interest in developing the land, or by a group such as the Nature Conservancy. For example, in November 1971 the Conservancy acquired 3,215 acres of Vermont mountainland which were being eyed by ski resort developers. The land will be transferred to Vermont at a cost to the State of less than two percent of the land's fair market value.

The Conservancy is not as well known as the Sierra Club or the EDF. It does not seek headlines or institute law suits or sponsor hikes through the wilderness or even publish nature books. But it has bought for posterity islands along the Atlantic coast, redwood forests, wild marshes and mountains and prevented developers from gobbling up thousands of acres in 44 states and the Virgin Islands.

The Conservancy buys the land itself, or lends money to groups dedicated to keeping the land natural. Much of its land has been donated by wealthy persons or corporations in the form of bequests and donations. It has been involved in the preservation of about 240,000 acres since its inception in 1917--most of it since 1960. In 1969 the Conservancy bought or received in gifts undeveloped land valued at almost $20 million. Part of its resources come from a $6 million line of credit granted by the Ford Foundation. Much of its strength comes from wealthy backers such as Lawrence Rockefeller, Arthur Godfrey and Marshall Field. In terms of actual expenditure and long-range effect, the Conservancy stands high among environmental groups.

SCIENTISTS' INSTITUTE FOR PUBLIC INFORMATION
Starting in 1958 from a few groups of concerned scientists and citizens, the Scientists' Institute for Public Information (SIPI), whose best-known members are Doctors Margaret Mead and Barry Commoner, has grown into a nationwide network of thirteen well-established scientists'

information groups. These groups attempt to provide "objective, reliable, scientific information to the technically untrained public." Most of the issues dealt with by SIPI and its affiliated scientists are now environmental in nature. The St. Louis affiliate of SIPI puts out an excellent periodical called Environment. This and all other local affiliates also issue reports on particular subjects such as mercury, air pollution and lead poisoning.

Science information groups operate by sending speakers to organizations requiring technical information and by providing information to the mass media. Members appear on radio and T.V. programs and work with other public interest groups who lack scientific expertise.

The New York-based science information group operated on a budget of less than $10,000 in 1970-71. Most of the technical work is donated by member scientists. A distinguished Board of Advisors (which includes Detlev Bronk, Rene Dubos, Margaret Mead and Gerard Piel) gives occasional advice to the working scientists.

The essential work of these groups is carried on by individual subcommittees. Some of the more active ones of the N.Y. group are those on air pollution, electric power and the environment, noise, water pollution, and lead poisoning in slum children. Changing times have caused the demise of the "peace science" subcommittee and the emergence of a transportation subcommittee. This latter group should expand rapidly as scientists and society begin to realize the key role transportation and fuel for cars plays in many pollution problems.

Scientist information groups have become widely known as sources for technical data to people concerned about environmental problems. SIPI scientists appear less susceptible to frenzy than many other environmentalists, and have earned a well-deserved reputation for impartiality and a dedication to truth that is often lacking in scientists employed by private companies and government agencies.

THE SIERRA CLUB

Founded in 1892 by naturalist John Muir to protect the High Sierras, the Sierra Club has grown to be one of the most vociferous champions of environmental causes that relate to wilderness. The club's statement of purpose opens with the following: "To explore, enjoy and preserve the Sierra Nevada and other scenic resources of the United States." Primarily a California organization of outdoorsmen with most of its more than 140,000 members west of the Rockies, it has in recent years changed its emphasis considerably to include other natural areas and broader pollution issues.

On joining the club in 1970, I received a form letter from Edgar Weyburn, M.D., Vice-President, which neatly sets the militant tone of the club. In part the letter reads:

Dear Member:

Welcome to the Sierra Club in its 78th year.

You are now a member of a unique organization; an outing club, an educational and scientific organization, a publishing house, and a conservation army. We are maligned and admired simultaneously, and you will shortly discover that while you may have to stand up for the choice you have made in joining the Club, you will not have to apologize for it.

The Sierra Club is the largest outdoor club in the nation with the broadest spectrum of outdoor activities. Members will be involved this year in mountaineering, rock climbing, hiking, downhill and across-country skiing and river running. Some 150 Club-sponsored outings will take in back country from Alaska to Peru, and thousands of members will go on chapter and group hikes and trips...

The Sierra Club has led the fight to preserve the coastal redwoods, the Grand Canyon, the North Cascades; we have fought for the security of wilderness areas in National Parks and Forests. Our 30 chapters have tackled local conservation problems...cutting off threats before they have grown to national size...and have provided grassroots leadership in preserving the quality of the human environment.

We will be battling today and tomorrow...

For $17 (or more) a year, a Sierra Club member gets the colorful monthly, Sierra Club Bulletin, and a newsletter describing local activities such as canoeing and hiking trips organized by the club. The club operates a thriving book-selling business that currently offers over 40 titles ranging in price from $2 to $30 each. Members are also welcome to come to local meetings and become active in organizational affairs and environmental projects

A good deal of Sierra Club money and energy goes into outings in the wilderness, ski trips, river running and general communing with nature by its relatively affluent membership. The club maintains headquarters in San Francisco and lodges and properties in the mountains of California.

The club was the principal sponsor of the 1964 Wilderness Act which established a National Wilderness Preservation System, a system that started with nine million acres but aims to add another 50 million acres by 1974. During the past few years the club has been extremely active in attempts to ban DDT, in battles to stop the Alaska pipeline and to eliminate clear cutting by timber interests and other non-recreational use of national forests. These topics are dealt with in greater detail in other portions of this book.

ENVIRONMENTAL DEFENSE FUND

"Sue the Bastards" is the unofficial motto of the Environmental Defense Fund (EDF), a vociferous group whose members are less restrained than those in SIPI. Two EDF scientists, doctors Charles Wurster and Max Blumer, have been in the forefront of public relation and legal battles to ban DDT and prevent oil spills, respectively.

One can sense the spirit of the EDF from the opening paragraph of a letter from Roderick Cameron to prospective members:

Dear Friend:

I hardly need to tell you about pollution! Your eyes smart from smog. Your nose is assaulted by factory fumes. Pesticides contaminate your children's food

64

and your ears ring from the roar of traffic and machines. You watch helplessly as still another precious wilderness disappears in the name of progress and profit.

EDF publishes a newsletter that deals primarily with its latest litigious activities. For instance, the July 1971 newsletter listed lawsuits to stop DDT discharge into the Pacific, a suit for injunctions against the Four Corners power project in the southwest, petitions to stop power delivery to a pulp mill, a brief filed with the Supreme Court to prevent Walt Disney Productions from developing a resort in the Sequoia National Forest, a report on the EDF action that helped stop the completion of the Cross-Florida Barge Canal, and a case to prevent the use of phosphate-based detergents.

As of the fall of 1971, EDF had "four offices containing 23 full-time employees including eight attorneys and five Ph.D. scientists...EDF's Scientist Advisory Committee now numbers nearly 700, and the Legal Advisory Committee contains about 60 attorneys..." supported mainly by 9,000 dues paying members. EDF is involved in approximately 50 environmental cases.

EDF often seems to operate more in an evangelical than a scientific spirit. Having once made up their collective mind that DDT should be banned or that some other environmental issue is worth fighting for or against, EDFers resort to almost any tactic to win the fight. In a letter to the New York Times (August 18, 1971), Roderick Cameron, Executive Director of EDF, accused Dr. Thomas Jukes, a controversal proponent of moderate DDT use, of being a tool of the pesticide industry and of having published nothing on the subject of pesticides. Dr. Jukes is a distinguished scholar and a professor at the University of California (see p. 331.)

EDF may instigate legal battles to gain public attention for its point of view, so that in "suing the bastards" its aim may simply be to gain publicity even when it knows its arguments are not justifiable and that it will lose the case.

COUNCIL ON ECONOMIC PRIORITIES

The Council on Economic Priorities (CEP) is a non-profit research group that concerns itself with environmental and social activities of individual companies. CEP consists of relatively young women and men with degrees in economics, business administration and law who seek to determine the social responsibility of corporations. They write well-researched reports dealing with corporations' behavior on such issues as military involvement, dealings with repressive regimes, minority hiring and pollution control. The reports are sold to subscribers Adverse reports have depressed the stock price of some notable companies and tend to induce offending companies to take corrective action.

A set of reports on petroleum companies (August 1970) related their expenditures for pollution control and their history with regard to pollution as well as other social issues. The CEP research dealing directly with industrial pollution resulted in a 400-page report, "Paper Profits," that dealt with 24 major pulp and paper producers and their 131 mills. The study found that less than half of the paper mills provided adequate control of the "foul and pervading stench" associated with the paper making process. The study noted that the companies used over 2.3 billion gallons of water a day, and discharged more than a third of it untreated directly into lakes and streams.

By pointing the finger of environmental unconcern directly at corporations such as St. Regis paper, Potlatch, and Diamond International, and backing up charges with facts, CEP has been able to nudge these and other firms into more vigorous anti-pollution action.

In the past 10 years thousands of other environmental groups have sprung up to fight particular issues. The "Save the Dunes Council" is dedicated to the establishment of the Indiana Dunes National Lakeshore. It has 3,000 members. The "Society for the Preservation of Birds of Prey" is a private international organization dedicated to protection of eagles, hawks and other birds of prey. "Citizens for Clean Air" is a New York and New Jersey group advocating air pollution abatement in the metropolitan area. Ad Infinitum.

The thousands of environmental groups include well-meaning individuals who for a variety of motives are concerned about pollution, ecology and the deterioration of the natural environment   This concern has helped educate millions of Americans who otherwise would know little about the complex issues involved in ecology, and spur corrective legislation.

THE LEGAL BATTLE AGAINST POLLUTION

As polluters are usually not willing to cease polluting unless forced to, the forces seeking a cleaner environment have resorted to legal action. A body of new laws attendant with a new breed of environmental lawyer and enforcement agencies, has sprung into existence during the late 1960's.

The two main laws for forcing compliance of industrial polluters are the Refuse Act of 1899, which relates to water pollution, and the Clean Air Laws of 1970.

The Refuse Act of 1899 prohibits discharge of "refuse matter of any kind or description" other than sewage into navigable waterways and is (or was) under the jurisdiction of the Army Corps of Engineers, the agency charged with maintaining navigable waterways. The law lay practically dormant until the late 1960's when the furor about pollution came to a head. About 40,000 U.S. firms must apply to the Corps of Engineers for permits to discharge effluents into rivers, lakes, estuaries and the sea.

Under the most recent interpretations of the law, industrial firms were supposed to apply by July 1, 1971 to the Corps of Engineers for permits to discharge effluents. As of late 1972 many firms had not complied. Firms require certification by the state water pollution control agency that they are either already complying with federal standards, or that they are carrying out an approved program that will lead to compliance. Penalties for violations include fines up to $2,500 per day and up to one year imprisonment for each count. A feature of the law that has caused much industrial howling is that private citizens can receive a bounty for turning in pollutors.

Since 1965, the government has had more concern over the use and abuse of U.S. waterways. The Water

Quality Act of 1965 required the states to establish and enforce water quality standards for all intrastate waters. If the states failed to take proper action by June 30, 1967, the Environmental Protection Agency (EPA) is authorized to set Federal standards.

In 1966, the Clean Waters Restoration Act provided substantial funds to help communities pay for water treatment facilities.

The Water Quality Improvement Act of 1970 is concerned primarily with minimizing ocean oil spills.

In November 1971, the U.S. Senate approved a bill to strengthen the waste control program supervised by EPA. The bill, designed to stop all waste pollution by 1985, was passed by an 86 to 0 vote. Despite a veto by President Nixon, the Senate and House voted 52 to 12 and 247 to 23 on October 18, 1972 to pass the Federal Water Pollution Act of 1972, which authorizes the spending of $24.6 billion over three years to clean up the nation's rivers and lakes.

Under the Clean Air Amendment of 1970, firms have a three-year deadline in which to comply with abatement schedules but meanwhile they must monitor emissions and make regular reports to EPA.

By recently enacted law, the federal EPA administrator has authority to close down an industrial plant if deemed necessary to prevent pollution. The federal government is prohibited from entering into contracts with any violator of the Clean Air laws. Violations are punishable by fines as high as $25,000 per day and up to one year imprisonment. Federal authorities will have access to any plant suspected of polluting. Private citizens and environmental groups may file suit against alleged polluters. Private citizens may also file suit against government administrators who do not carry out their responsibilities in enforcing provisions of the Act

The National Environmental Policy Act (NEPA) of 1969 is the major statuatory lever to ensure environmental quality in Federal Government actions. Under NEPA, Section 201, promoters of any project that may have an effect on the environment must file an Environmental Impact Statement with EPA detailing all such possible effects. We will consider this procedure in detail shortly.

A host of other Federal laws bear on environmental matters. For example, the Department of Transportation Act of 1966 protects public parks and historical sites; the Fish and Wildlife Act of 1958 declares that wildlife be given "equal consideration" with other water uses; and the Wilderness Act and the National Park Service Act provide for preservation of wilderness and parks. The courts have extended the usefulness of these laws in favor of environmental protection and enforcement.

Starting in November 1971, federal prosecutors, in an attempt to get action against persistent pollutors, began to jail owners and executives of companies that discharge polluting wastes into waterways. The drastic step of prosecuting polluters as criminals rather than as civil offenders should speed up compliance.

A slew of environmental laws were enacted by Congress prior to the elections of November, 1972. President Nixon signed into law:

The Coastal Zone Act, the nation's first "land use" legislation aimed at protecting ocean and Great Lakes coastal regions. The coastal zone bill was originally proposed in 1969 by Senator Warren G. Magnuson. States will be given federal assistance in planning and financing programs to preserve coastal areas from encroachments by builders, population pressures and inadequate planning. The bill authorizes the spending of $200 million over a five year period.

The Ocean Dumping Act which, when it takes effect in April 1973, will ban the dumping of toxic and dangerous materials into the ocean. The law will establish a strict permit system for ocean disposal of sludges and spoils and other refuse which will not be harmful to the marine environment.

The Marine Mammal Act bans the killing of whales, seals, sea otters, porpoises and other species of seagoing mammals by citizens of the United States or in waters controlled by the United States. The law also bans import of products made from marine animals.

In the closing days of the 92nd Congress, both the House and Senate overrode a Presidential veto of the Federal Water Pollution Control Amendments and authorized roughly $25 billion to be spent during the next three years for water treatment plants. The new bill changes the enforcement mechanism of water pollution control programs from water quality standards to effluent limitations.

ENVIRONMENTAL IMPACT HEARINGS
AND SITING PROBLEMS

Because human ecology is a relatively new science and pollution monitoring and abatement technology is a young and often complex engineering discipline, laymen often take science fiction for science fact in this area. In the United States, Environmental Impact Hearings sponsored by all levels of government have become a form of national circus where any citizen, no matter how misinformed, can get up and say his piece regarding the project at hand. To site a power plant often requires 40 hearings and reviews. At each open hearing, the group trying to clear a particular location must defend its choice and usually employs a battery of experts who testify one way; while the environmental groups opposing the site employ a battery of experts who testify another way. Environmental impact proceedings have already held up the building of dams, pipelines, power plants and other badly needed public facilities.

Siting new U.S. electrical power plants has become a particularly difficult task; in many areas of the country it has become almost impossible. The nation's need for more electric power is growing faster than our capacity to generate it, as recent brownouts and blackouts have shown. One obvious answer would seem to be to build more power plants, but an alternate valid solution is to reduce the need for electrical energy. One may well question the assumption that doubling power capacity every 10 years is either necessary or wise.

Though Houston, a rapidly growing city, lacks adequate electric power, environmentalists won't allow the local power company to start building new generating plants because the effluents, even when cooled, would raise the temperature of the receiving waters some two degrees above the temperature that would support present marine life.

From an air pollution abatement viewpoint, nuclear or fossil power plants away from urban areas are more desirable than fossil fuel plants in cities. However, the cost of power transmission lines often make remote sites uneconomic.

Environmentalists have been instrumental in preventing Consolidated Edison from building a pumped storage, two-million kilowatt power plant at Storm King Mountain on the western shore of the Hudson River, north of New York City. The Cornwall plant, originally scheduled to go into operation in 1967, is presently being contested in the courts. It has been approved twice by the Federal Power Commission. In November 1971, a three-man court again approved an FPC license for Con Ed. The majority of the court noted that the Storm King project would be the largest pumped storage plant in the world and could provide enough reserve power to prevent future blackouts. The court concluded that "there was no satisfactory alternative" to the power plant and that "the scenic impact would be minimal." Despite this ruling, conservationists headed by the Scenic Hudson Preservation Conference have asked for a hearing before the United States Court of Appeals and have vowed to take the cause to the U.S. Supreme Court, if necessary.

Perhaps the most far-reaching decision yet reached as the result of an environmental impact hearing was the Calvert Cliffs case, involving three nuclear power plants in Maryland, on the Chesapeake Bay. The decision of the U.S. Circuit Court of Appeals for the District of Columbia Circuit in Calvert Cliffs Coordinating Committee v. Atomic Energy Commission (Nos. 24839 and 24871, July 23, 1971) will stand as a basic precedent for future interpretations of the National Environmental Policy Act, as well as of other state and federal environmental legislation, both present and future. This first decision of a federal appellate court construing Section 10A of NEPA applies specifically to the Atomic Energy Commission, but it has broad implications for the environmental impact analysis of all federal government actions subject to NEPA. The decision also has broad implications for industry and the states.

Environmentalists had long contended that the AEC was not carrying out its responsibilities under NEPA. The Court in deciding that the AEC was shirking its responsibilities interpreted key parts of the Act, namely Sections 101 and 102. Judge J. Skelly Wright wrote, "These cases are only the beginning of what promises to become a flood of new litigation – litigation seeking judicial assistance in protecting our natural environment. Several recently enacted statutes attest to the commitment of the government to control, at long last, the destructive engine of material progress. But it remains to be seen whether the promise of this legislation will become a reality. Therein lies the judicial role... Our duty, in short, is to see that important legislative purposes, headed in the halls of Congress, are not lost or misdirected in the vast hallways of the Federal bureaucracy...that the vagueness of the NEPA mandate and delegation (of authority) leaves much room for discretion. We find the policies embodied in NEPA to be a good deal clearer and more demanding than does the Commission." Judge Wright draws a clear distinction between the Section 101 language, which states that the federal government must "use all practicable means and measures" to protect environmental values, and the "to the fullest extent possible" language of Section 102. "We must stress as forcefully as possible that this language does not provide an escape hatch for footdragging agencies; it does not make NEPA's procedural requirements somehow 'discretionary.'"

As a result of the Calvert Cliffs decision, the AEC has decided to referee nuclear questions, and no longer promote nuclear power, as had been the case. AEC Chairman James R. Schlesinger, in a blunt talk to a gathering of nuclear industry people at a banquet in Bal Harbour, Florida, advised them to take their case to the public as does the Sierra Club. Mr. Schlesinger went on to say:

Environmentalists have raised many legitimate questions. A number have bad manners, but I believe that broadside diatribes against environmentalists to be not only in bad taste, but wrong... The question had been raised by Michael McCloskey of the Sierra Club among others, whether our society, for environmental reasons viewed broadly, ought not to curb its appetite for energy and for electric power. It is a legitimate social question. It is not unreasonable to question whether neon signs or even air conditioning are essential ingredients in the American way of life. More fundamentally it is not unthinkable to inquire whether energy production should be determined solely in response to market demand.

To avoid siting problems some firms are consulting with environmental groups before choosing sites. The Northern States Power Company called in local environmental groups to help them site a power plant near Milwaukee. The group came up with a site that was not the company's first choice, but which the company nevertheless accepted and thus avoided a costly and time-consuming court battle.

Delays in obtaining nuclear power plant licenses are expensive. Licensing action on over 100 nuclear plants with a total capacity of 90,000 megawatts has stopped until the Atomic Energy Commission reviews the environmental reports required from utilities under the agency's new regulations. An executive of the Baltimore Gas and Electric Company has estimated that a one year delay for 55 of the plants affected by the AEC's NEPA provisions would cost utilities from $5 billion to $6 billion. These costs will undoubtedly be passed on to customers in the form of higher rates.

I was involved in a minor way with an impact hearing in North Hempstead, a town in Nassau County. Roslyn Village, a wealthy suburb on Long Island, borders the cod end of Hempstead Harbor. Roslyn boasts a modern incinerator that burns the garbage (euphemistically referred to as solid waste) of North Hempstead Township, of which Roslyn is one of the smaller villages. Only when the Christmas trees are burned does Roslyn smell sweetly.

Since 1900, the wetlands on the North Shore of Nassau County have shrunk from 380 to 140 acres; in Hempstead Harbor the number has declined from 165 to 35 acres.

The town of North Hempstead would like to fill and use another 26 acres of shoreline in the harbor, including eight acres of wetlands. The town wants to dump incinerator residues on the filled-in land.

Hempstead must get a permit from the Corps of Engineers in order to fill in the wetlands. Among the interested parties who attended a public hearing on the issue were the Roslyn Chamber of Commerce, Citizens for a More Beautiful Port Washington, the Long Island Environmental Council, the Roslyn Ecology Association, and other groups who would stop any further changes of the harbor. The preservationist groups presented a negative report from the U S. Fish and Wildlife Service, stating that the eight acres that would be lost are an invaluable ecological asset.

Leaders of anti-fill groups say that the town's remedy is a poor stopgap until the county comes up with a recycling program that will once and for all solve the problem of solid waste disposal.

The town commissioner of public works, after a good deal of study, concluded that the town needed additional space in which to put the burned residue from its garbage incinerator. The marshland area had been decided on many years before when the incinerator was first built and included approximately eight acres of unsightly marshland in Hempstead Harbor. At least a dozen agencies became involved in this project, including EPA, the Army Corps of Engineers, the state health agencies, the fish and wildlife agencies, county agencies of all shapes and sizes, town agencies, and agencies and representatives from the surrounding villages.

On a nasty, rainy February night, hundreds of people showed up for the hearing, which was held under the auspices of the Army Corps of Engineers. The meeting went on until 1 AM as scores of opposing "experts" and citizens testified on the advisability or desirability of constructing the dike in Hempstead Harbor. Most of the local speakers decried the choice of site. Some "environmentalists" recommended using local sand pits instead, although it was pointed out that this would not only be extremely expensive, but also could very well poison the local ground water supply.

I was asked to help prepare an "environmental impact" statement on the possible effects of the proposed plan of North Hempstead to extend the dike. My testimony involved the question of whether or not marshlands had been created in the vicinity of the existing incinerator dike, and concerned the possibility of creating new marshlands in the harbor. The opening paragraph of my statement was:

> We humans are strange animals; only a few years ago salt marshes were regarded by most people as nuisances, eyesores and places where mosquitoes bred. We applauded the destruction of marshlands. Now some of us realize that wetlands are the breeding

grounds of most of our local finfish and provide shelter and nourishment for millions of dollars worth of shellfish and finfish, and that marsh plants help purify estuarine waters.

The issue of the incinerator refuse dike is still unsettled. The complexity of this relatively small project, proposed by a small town to improve the environment by getting rid of solid waste in a sanitary manner, gives some idea of what is involved in larger projects. Construction costs in the Roslyn area are rising at a rate of over 1 percent per month so that each month's delay in the combined $16 million embankment extension and plant expansion project will cost the town over $160,000, or approximately one dollar per person per month. Use of the alternate sand pits as a fill site, if it were ever allowed by the local health authorities, would cost, according to Town Supervisor Michael Tully, an extra $4 million. And as the impact hearings drag on, Hempstead's garbage piles up.

Environmental Impact Statements may develop into voluminous documents to which more than a hundred persons have contributed. With the newly-spawned crop of environmental experts in the land it is virtually impossible to prepare a major statement that couldn't be proven somewhat deficient. Judges, presumably trained in the law, must base their judgement on conflicting testimony of "experts." Often the simplest recourse is to do nothing, and in that way stop a proposed project.

The case of the Dusky Seaside Sparrow illustrates what effect Environmental Impact Statements may have. Agencies in Florida had planned a major highway out of Jacksonville, when a local bird fancier discovered that the proposed highway would go through an area that might be a suitable habitat for the Dusky Seaside Sparrow. It was uncertain whether these sparrows actually inhabited this area since no one claims to have seen them there. Though they had been spotted some distance away, environmentalists could not prove that Dusky Seaside Sparrows lived in the area or establish that the road would destroy their habitat if they did; but because of the possible harm to the Dusky Seaside Sparrow, the road route was changed at a cost to the taxpayer of some $2 million.

On my way to jury duty last October, I noticed four well-dressed matrons parading in front of the Federal Court House in Foley Square, carrying large signboards reading, "Protect Our Environment." I stopped to talk with one of the ladies, and she told me that she felt her environment was in danger because the Federal Government intended to build a badly-needed nine-story jailhouse as an annex to the existing Courthouse. Residents of the nearby Chatham Square area had invoked the NEPA act in a so-far successful attempt to stop construction of the rather small jailhouse. It is unlikely that the framers of the original legislation intended that the NEPA act be put to such use.

THE COST OF CLEANING UP

As with most economic problems, the individual and society must make a choice between ridding the environment of a certain pollutant at a certain cost, or spending the equivalent money elsewhere. The cleanup cost estimates are staggering.

The environmental movement is the first mass movement to be based primarily on aesthetic values and concern for nature rather than economic or religious beliefs. Nevertheless, basic economics underlie attempts to clean up the environment. What does pollution cost society? It is hazardous, perhaps impossible to estimate. Evidence is rudimentary and conflicting as to health hazards that are usually said to be the highest cost of pollution.

The White House Council on Environmental Quality (CEQ), in its second annual report (August 1971), estimated an annual cost of $6 billion to human mortality and morbidity resulting from pollution, but as we shall see, the links have not yet been forged between air pollutants and health hazards. And as for deaths or disease from polluted water supplies, Dr Abel Wolman, Professor Emeritus of Sanitary Engineering at Johns Hopkins, recently observed that fewer people are dying in the U.S. of water-borne disease than ever before.

Water-borne diseases have been near the vanishing point for many years. Despite the variety of chemicals now reaching raw water sources (DDT, mercury, phenols, and so forth), almost no evidence exists to show that they are a menace to health, although one would not know this by reading the headlines in most periodical articles written on the subject. In a talk before the Fifth International Water Quality Symposium in early 1971, Dr. Wolman called attention to the 2 billion people in other countries whose water supplies are unsafe. Some 10 million people in poorer lands die annually because of hazardous water, but not in America, the world's most highly industrialized nation. Yet the CEQ report ascribes $6 billion to human mortality and suffering from pollution of air and water.

Other lesser costs estimated by CEQ are $4.9 billion in damage to crops, plants, trees and material, and $5.2 billion in lowered property values. These latter figures, though probably more reliable than the one on health costs, are extremely speculative; suffice it to say pollution causes billions of dollars worth of damage.

The CEQ in its second annual report, 1971, estimates that it will cost $105.2 billion over the next six years to "clean up" the American environment. Of this total, $23.7 billion would go for cleaning the air; $43.5 billion for solid waste control, and $38 billion for water.

About two-thirds of water pollution costs must be borne directly by government. Most of this $26 billion dollars would go for sewage plants and sewers. Almost all of the air pollution control costs would come from private industry.

An "improved" cost estimate appeared in the August 1972 report of the CEQ, in which the council opined that it would cost $287 billion to solve the major pollution problems of the United States. Air pollution control will require $106.8 billion, and water pollution control will cost some $87.3 billion, according to CEQ economists. Spread over a decade, annual expenditures are slated to rise from $10.4 billion in 1970 to $33.3 billion in 1980. The money would come from private as well as public sources.

According to an industrial survey done by the McGraw-Hill Publications Department of Economics, the industrial pollution fight alone will cost about $18 billion. Presently American industry claims that it is spending about $3.6 billion a year in an attempt to satisfy by the end of 1975 current anti-pollution legislation. But since standards, such as those for automobile exhausts, sulfurated fuels and water quality, are likely to become more stringent by 1975, the total amount required may blossom to much more than $18 billion.

According to the McGraw-Hill report, the total amounts required to bring the major industries up to standards are:

Electric power	$3.24 billion
Iron and steel	$2.64 billion
Petroleum	$2.12 billion
Paper	$1.89 billion
Chemicals	$1.00 billion

During 1971, expenditures for air pollution accounted for $2 billion of the total, and for water pollution control, $1.6 billion. Since these are figures given by industry, they may be somewhat biased, but they do indicate the magnitude of current anti-pollution expenditures.

Air and water are peculiar commodities in that they belong to everyone and no one; they move across national boundaries; they envelop us — and up to a few years ago, we took them for granted. "Certain commodities such as air and water," said H. H. Houthhakker in April, 1971, "were free until recently because the supply vastly exceeded the demand. Since they were free, or virtually free, there was no need to establish property rights or other procedures for deciding on their use. As demand has grown, however, the supply has become inadequate and we therefore have to find ways of allocating these scarce commodities among different uses, that is to say, of using them efficiently. Because of their mobility, it is much more difficult to attach property rights to air and water than to land. Indeed even in the case of land, the mere establishment of property rights is not necessarily sufficient to obtain the best use."

Society can pay for controlling pollutants by setting legal standards and forcing industry to meet them, or by taxing polluters and giving tax abatements for pollution equipment. Either way, society must pay in higher prices for products, or in loss of tax revenue. It is noteworthy that the cost of removing pollutants from air or water rises rapidly as standards increase. It costs $1 per pound to remove 30 percent of the biological oxygen demand (BOD) from waste water; $25 per pound to remove 60 percent of the BOD; and $60 per pound to remove 95 percent of the BOD. Total (100 percent) removal of pollutants from water costs five times more than 85 percent removal. A similar disproportionate increase in abatement costs holds true in removal of pollutants from air.

Because local communities may set emission standards higher than those suggested by the Federal Government, some industries will be forced by simple economics to leave a state such as New Jersey, for a less regulated one such as Arkansas. Or an industry using large quantities of water may simply move to a location where the water is more abundant. Therefore it is imperative that pollution standards be set uniformly on a national basis. Hopefully, international standards will become uniform.

In his letter of transmittal of the 1971 CEQ report to Congress, President Nixon said:

> Our efforts will be more effective if we approach the challenge of the environment with a strong sense of realism. How clean is clean enough can only be answered in terms of how much we are willing to pay and how soon we seek success.

"How Clean Is Clean Enough" would make a good subtitle to this book. Despite strong personal disagreement with the President regarding most issues, I find myself in complete agreement with his statement. I note, however, that the Nixon administration has done little but talk so far regarding environmental cleanup. Most of the government agencies assigned the task of improving the environment have been virtually starved for funds. In August

1971 it was pointed out in the <u>New York Times</u> that Administrator William Ruckelshaus of EPA had to water down his statement regarding pollution standards because of pressure from the White House. Nixon's ex-Secretary of Commerce, Maurice Stans, has been particularly vocal in his suggestions to "go slowly" on pollution issues. Copper companies were able to persuade the Federal Government to lower sulfur emissions standards for smelters.

Since over $100 billion dollars is involved and the fate of many marginal companies and areas hangs in the balance over the pollution question, it is not surprising that politicians and businessmen have taken a lively interest in the environmental debates.

"ECO-PORNOGRAPHY"

Mass media advertising is the most powerful sales tool ever devised for inducing people to buy on a national scale. Advertising often creates desires and attitudes that lead to heedless consumption. Millions of persons view ads placed in the mass circulation periodicals, or broadcast on television. Madison Avenue has played an integral role in shaping modern life and life styles that affect the environment. Advertisers have sensed the public concern about ecology and some unscrupulous advertisers have sought to capitalize on this concern in their ads.

Jerry Mander, a San Francisco Advertising executive and member of the Sierra Club, is credited with coining the word eco-pornography to describe advertisements pandering to the public fear of pollution and the widespread desire for a better environment Advertisements touching on ecological issues are often technically, and sometimes deliberately, misleading. Mander's advertising agency was summarily fired by an automobile agency after Mander publicly raised questions about the desirability of the internal combustion engine. Mander shares the opinion with Rene Dubos and others that the U.S. must create a stable "no-growth" economy in order to deal with major pollution problems; an opinion that is an abomination to aggressive industrial executives and the advertisers who pander to them.

Ecologically destructive ads come in many different forms. Ads such as those for automobiles, unnecessary detergents, cigarettes and snowmobiles create desires (and use) of environmentally harmful products.

Ads may present only one side of an environmentally controversial issue and thus mislead the public. Humble Oil and Refining Company presented a slick series of films on the "Meet The Press" show relating to oil and Alaska. One film indicated that the country is heading for a fuel shortage and there is a desperate need for more oil. Other film clips showed that pipelines built in northern climates are safe, and that any destruction to tundra can be eradicated. This expensive film series was an attempt to persuade the American public to allow the oil industry to build the Trans-Alaska Pipeline (TAPS) in which Humble has more than a passing interest.

Other notable examples of eco-pornography are:

Ford and General Motors claim that their newly designed engines have eliminated 80 percent of the hydrocarbons in engine exhausts. The ads failed to point out that the above figures apply only to expertly tuned engines, and that 75 percent of all engines failed to meet these rigid standards.

AMOCO's advertisement of lead-free gas as a beneficial fuel, ecologically speaking, ignored the fact that surveys in Nassau and Suffolk counties in New York State found lead in Amoco gas ranging from .0147 to 2.8 grams per gallon in 24 gasoline stations.

Potlatch Forest Products, Inc., sponsored a full page color ad showing a bright blue river, flecked with foam as it runs over rocks with evergreen-covered hills on either shore. Underneath this scene of natural beauty was the caption: "It cost us a bundle but the Clearwater River still runs clear." What the company failed to mention was that the picture in the ad was taken 50 miles upstream from the company's pulp and paper mill in Lewiston, Idaho. In the vicinity of the plant some 2.5 million tons of sulfurous gasses foul the air each year and the river is polluted by 40 tons of organic wastes that the company dumps in the river each day.

An ad publicized by Southern California Edison shows a large lobster over the caption: "He likes our nuclear plant." Aside from the fact that the lobster showed no such affection, it turns out that the lobster was borrowed from a local marine biologist who expressed the opinion that the nuclear plant is harmful to the environment.

Standard Oil of California (Chevron) jumped on the eco-pornography bandwagon in January 1970 with announcements of a gasoline additive with the name F-310, "a major breakthrough in pollution control technology." The company has spent several million dollars in advertisements featuring ex-astronaut Scott Carpenter and advertisements showing a large balloon turning black with exhaust fumes from competing gasolines, while the exhausts are clear using "just six tankfuls of Chevron with F-310." These ads attempt to capitalize on the Californians' concern with air pollution. But the Federal Trade Commission (which is looking into a growing number of eco-pornographic ads) believes that Chevron is using a common detergent used by most other oil companies and that the black exhausts shown in its balloon ads were fabricated for their photogenic value.

Companies attempting to take advantage of public concern to further their own ends are only one example of socio-economic trends of the pollution wars. It would be ironic if industrial concerns, most of whom value profits above ecology, could through advertising create an image of themselves as saviors of the environment.

Costs of damage done by pollution and foolish consumption can be considered as "diseconomies" and in fact have been studied by a few "diseconomists" who have questioned the value of economic growth as measured by the gross national product (GNP). The GNP tells nothing of the pleasure a society may derive from its economic activities and there are those who argue that higher GNP's have led to lower quality of life, and less contentment. The destruction downwind caused by the smoke from an industrial plant, damage to plants and lungs, soiling of clothes, corrosion of painted and stone surfaces, are extremely hard to evaluate, but they are nevertheless real diseconomies. On a broader scale pollution of air, water and land are diseconomies that have accompanied the growth of industrial economies.

Thomas Nast 1880.   Harper's Weekly.

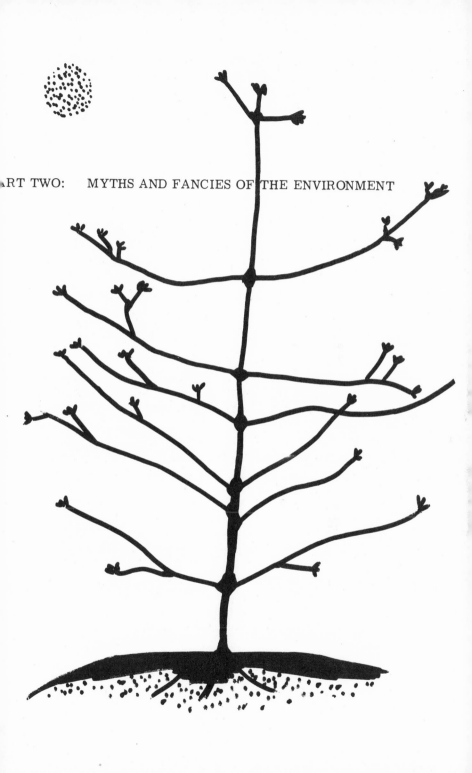

ART TWO: MYTHS AND FANCIES OF THE ENVIRONMENT

4

CITY AIR IS KILLING US

Thirty dirty birds
Were flying through the sky;
So we chased them and we chased them
Till they said "bye, bye."

Peter Agnos

"I am afraid to take a deep breath of air in New York City," said a woman I know, and there are thousands of people who feel as she does. An old psychiatrist friend asked me, "How can these people live normal and productive, let alone happy lives, without ever taking a deep breath?" It is a good question for many city dwellers in whom the fear of polluted air has become more hazardous and debilitating than the air itself.

The widespread myth of city air would have city dwellers believe that the air we breathe is becoming increasingly loaded with pollutants that are causing widespread disease and death. According to Rene Dubos, "We are running out of breathable air in many cities ." My favorite quotation is by the lawyer Michael Belknap, Director of the New York City Council on the Environment: "Although proof of the cause-and-effect of pollutants and disease is not established, the evidence is overwhelming."

Pollution of city air – that is, the addition to air of contaminants such as sulfur dioxide, carbon monoxide, lead particles, and oxides of nitrogen – does indeed exist. But is it a dangerous and worsening condition? To read some of the literature put out by anxious environmentalists, one gets the feeling that city dwellers are inexorably doomed by each inhalation. I would like to assure the city dweller that he still has a few years to live.

Pollution of city air can be divided into two major types: the Los Angeles oxidizing smog composed mainly of automobile exhausts, and the London reducing concoction composed mainly of fog, sulfur compounds and particles derived from the burning of coal and oil for space heating and power generation. New York, Northern European and Japanese cities tend to follow the London pattern. Most other American cities veer toward the Los Angeles type of environment. London itself is now free to a great extent from London-type air pollution because of changes in space heating patterns. Almost all of London is now heated by relatively low sulfur oil or natural gas from the North Sea.

THE POLLUTANTS

Industrial man has changed the air around him possibly for the worse. He has added nuclear fission, asbestos and lead particles, sulfur dioxide, nitrogen oxides and carbon monoxide, to name some of the most common and potentially injurous contaminants. But the trend over the past few years has been to reduce most of these contaminants in cities in the United States, Europe and the Soviet Union. We now find considerably reduced concentrations of radioactive particles in the atmosphere and thus over our cities by virtue of the banning of above-ground nuclear explosions. In a year or two it is likely that the Chinese and the French, having displayed their virility, will dispense with aboveground atomic explosions.

Nitrogen oxides (NOx) and hydrocarbons come mainly from the combustion of fossil fuels such as gasoline and coal. In the presence of sunlight, nitrogen oxides interact with hydrocarbons to cause photo-chemical oxidants that are eye, nose and throat irritants. In the U.S. about 15 million tons of hydrocarbons and about 7.5 million tons of oxides of nitrogen were generated in 1970, of which about 70 percent came from cars and almost all of the remainder from stationary power sources.

In our cities, sulfur dioxide (SO_2), an unpleasantly smelly pollutant, derives mainly from the burning of coal and oil and to a lesser extent from combustion of other fossil fuels and garbage. In New York City, electric utilities contributed about 40 percent of the 380,000 tons of sulfur dioxide generated in 1970. Home heaters and incinerators contributed most of the rest. In the atmosphere SO_2 can combine with water to form sulfuric acid, a corrosive pollutant.

Carbon monoxide (CO) is a colorless, odorless, tasteless gas. Almost all carbon monoxide comes from automobile exhausts, the percentage reaching 95 percent in Los Angeles. Carbon monoxide forms when a fossil fuel does not completely burn. Concentrations of greater than 20 parts per million can make one nauseous and indeed, as with most seemingly harmless substances, over-exposure to CO can cause death.

Particulate air pollution includes a wide variety of liquids and solids such as oil and mists, soil particles, grime and salt spray. Soot and dust in the city air can soil shirts and dresses. Particle sizes range from 50 microns * down to less than one micron. Most city air particles range from 1/10 micron to 10 microns. In New York City, private home and apartment house incinerators are still the main source of soot, although Local Law 14 requires upgrading or replacing of faulty incinerators. The small particulates in city air can be breathed into the lungs, and can contribute to reducing visibility.

Lead particles in city air are almost entirely derived from tetraethyl lead (TEL), put into gasoline to allow high compression automobile engines to run on lower octane fuel. The recently enacted ban on the use of lead additives in gasoline should reduce the amount of lead in the atmosphere and in our earth's waters.

Hydrocarbons are unburned chemicals in combustion, primarily from car exhaust, which react in air to produce smog.

CAUSES OF AIR POLLUTION IN CITIES

Man's inadvertent pollution of the air is still relatively insignificant compared to what nature has done and is likely to do to the earth's atmosphere Many times during the history of the earth, forest and grass fires caused by lightning or other natural accidents have wiped out millions of square miles of vegetation and filled the world's air with smoke and oxides of nitrogen and carbon. It is encouraging that since man's occupation of forests and

* A micron is one 25,000 of an inch

savannas, air pollution due to catastrophic continental forest fires has been greatly reduced.

Volcanic eruptions are another cause of natural air pollution that man must go a long way to match. In 1883, Krakatoa exploded and caused the sunsets around the world to glow red for many years afterwards because of the enormous quantity of particles blown into the atmosphere. That explosion could be heard 3,000 miles away, an example of noise pollution that has yet to be equaled by man.

More than 260 million tons of air contaminents are injected into the atmosphere of the United States each year. Of these 260 million tons, over 50 percent is due to transportation sources; 16 percent comes from power plants and other stationary sources; 15 percent derives from industrial processes; about 4 percent is from incineration of solid waste; and about 14 percent from fires in forests, on farms and in coal waste dumps. (See table 2-1, p. 44.)

Eighty percent of all U.S. air pollution comes from burning of fuels. This includes 55 percent of all carbon monoxide, 95 percent of all oxides of sulfur, and 85 percent of all oxides of nitrogen.

The major cause of most of the undesirable pollutants in the air over most large American cities is the automobile internal combustion engine. The automobile contributes some 43 percent of all pollutants by weight in the U.S. as a whole. Automotive exhausts contribute practically all of the lead, carbon monoxide, and oxides of nitrogen, plus a good deal of the sulfur dioxide. In Los Angeles, where roadways cover one-third of the city area, the California Department of Public Health, Bureau of Air Sanitation reports that automobiles contribute 89 percent of the total atmospheric pollutants, compared with 1.6 percent contributed by aircraft; while all other sources account for the remaining 9.4 percent   Included in the 89 percent of air contaminants derived from automobiles are 70 percent of all hydrocarbons, 95 percent of all carbon monoxide, and 65 percent of oxides of nitrogen.

Even though New York City has the largest, most versatile public transportation system of any American city, the automobile still contributes more than 75 percent (by weight) of all air contaminants to New York's air.

89

The remaining 25 percent or so comes from power generators that burn fossil fuel, space heaters, incinerators and industrial sources.

THE FIGHT TO REDUCE POLLUTANTS

In many cities improved rapid transit systems and a ban on unnecessary automotive travel would cause the air pollution problem to vanish with the winds. However, like cigarettes, automobiles are a persistent habit, especially since their sale is vigorously promoted by enormous advertising expenditures. For a variety of reasons, many Americans have scattered to the suburbs where they are at the mercy of their cars for basic transportation. Cars feed on suburbs and suburbs depend on cars. It is noteworthy that the 1.5 million residents of Manhattan own only 200,000 vehicles, or one car for every eight inhabitants. In suburban-sprawled Los Angeles, there are four million cars for seven million residents, more than one car for every two persons.

Under the Clean Air Act Amendments of 1970, the Environmental Protection Agency established national ambient air quality standards specifying the maximum levels to be permitted in the ambient air of the six principal and most widespread classes of air pollutants: particulate matter, sulfur oxides, hydrocarbons, carbon monoxide, photochemical oxidants, and nitrogen oxides. States are supposed to implement plans for limiting the emission of pollutants so as to achieve air quality standards by mid-1975. If any state should fail to develop or carry out such plans, EPA is authorized to do so. EPA also established standards limiting emissions from stationary sources of pollutants. The first such performance standards issued cover large steam-electric generating plants, municipal incinerators, cement factories, and sulfuric and nitric acid plants. Emission standards were established for new motor vehicles. Standards have been set requiring a reduction of 90 percent in hydrocarbons and carbon monoxide emitted by 1975 models as compared with the 1970 requirements, and a 90 percent reduction in oxides of nitrogen by 1976.

The EPA may regulate or even prohibit the manufacture or sale of fuels or fuel additives that result in harmful emissions or interfere with motor vehicle pollution control devices. The first such regulation will cover alkyl lead.

Incentives are provided to encourage the development of low-polluting motor vehicle propulsion systems, including government purchase and use of vehicles employing such systems. California and New York have similar programs. Citizens are specifically authorized to take civil court action against private or governmental officials failing to carry out the provisions of the law. Public hearings are required at various steps in the standards-setting, enforcement, and regulatory procedures to enable all interested persons to make their feelings known.

Lead particles in the air are almost entirely due to vehicular fuels. Airborne lead has been cited as a potential hazard to health, although it has not been shown to have caused injury to human beings living in cities. A 1971 report of the National Academy of Sciences was equivocal on this point. Nevertheless, it is heartening to note that once scientists established leaded gasoline as the cause of most airborne lead, society took steps to minimize the problem by reducing lead levels in gasoline. Despite the lack of hard evidence that lead in air is injurious to health, it is difficult to disagree with the ban of leaded gasoline because all the effects of lead on animals or plants that we know about are deleterious; there are no obvious beneficial effects of lead. Decrease in auto travel coupled with non-leaded gasoline must bring about decreased concentrations of lead in city air.

Unfortunately, most present-day gasoline engines are designed to run on leaded fuel They operate inefficiently on lead-free fuel and appear to emit greater quantities of oxides of nitrogen when lead is removed. Automobile manufacturers say that the problem of designing engines for non-leaded fuels may be solved in a relatively few years. The oil companies claim that the switch to lead-free gas will cost them $10 billion.

Much effort has gone into developing a "clean" engine for car use, and hopes have been raised a number of times in recent years for a pollution-free power plant. Each

91

great clean hope generated a lot of hoopla, but as yet we are still stuck with the internal combustion engine. A few years ago government groups heralded first the electric car and then the steam-powered auto as cures for the pollution problem, only to see the hopes fade in the light of harsh technical and economic reality. In September, 1971, the Environmental Protection Agency announced an experimental engine that they claimed "represents a breakthrough in emissions control technology and means that a truly clean car is not as far away as many people thought." The Ford Motor Company, which is developing the so-called "stratified charge" engine, immediately said that EPA's claims were exaggerated and said further that the engine, if indeed it ever could be perfected, was many years from production.

Is a practical engine that will not pollute available today? No. Will one be available in 1975 to meet the Government's air pollution deadline? In January of 1972, a Committee of the National Academy of Science (the high priesthood of American Science) reported that, "The technology necessary to meet the requirements for the clean air amendments for the 1975 model-year light-duty motor vehicles is not available at this time." The committee went on to say that they thought it "possible that the larger manufacturers will be able to produce vehicles that will qualify." But, the engines, if they are ever produced, will cost at least $200 more than 1973 models and will increase fuel consumption by 3 percent to 12 percent. Maintenance costs will also rise and, in addition, the cars will not drive as well.

There are two ways to reduce the pollution from automobiles: change the cars, or change the way they're used. Reducing the size of the engine and attaching pollution control devices will reduce the emissions for a given number of passenger miles. It would make more sense to reduce the number of passenger miles driven by improving mass transit facilities and penalizing users of cars in major cities.

We all would like our air to be cleaner, but it may be difficult to meet some of the requirements promulgated by the federal government to bring about this desirable end.

Henry Ford II noted that it would be "an impossible task" to meet federal standards for 1975. In an address given in Detroit on May 27, 1971, Ford said, "There is no way to complete a massive expansion of public transportation capacity throughout the country in the brief time between now and 1975. Nor is there any way in which public transportation can effectively serve most travel needs in metropolitan areas where jobs and homes have become more widely dispersed in response to the mobility provided by the automobile." Mr. Ford also said that the air quality goals themselves "are based on very questionable evidence" that the fumes are harmful. More will be said about that later in this chapter "Thus, we are faced by legal standards which demand enormous effort and expenditure to achieve goals for which there is little or no demonstrated need."

If they work, U.S. air pollution laws are expected to produce a decrease in the total automobile emission from 1970 until about 1980 (expressed in tons in whole atmosphere), but with a simultaneous continuous increase in oxides of nitrogen within that total. But, if present trends in auto use continue, during the 1980's all three auto emission pollutants (nitrogen oxides, hydrocarbons, carbon monoxide) will rise to a total of 55.3 million tons (39 million of carbon monoxide). Without this legislation emissions would have reached 187 million tons by 1990.

Citizens for Clean Air (CCA) called a public meeting in New York City on January 5, 1972, to discuss state implementation of federal air pollution standards. Peter Kahn, a spokesman for CCA, said that it would be virtually impossible for the authorities to meet the 1975 standards set for particulates in the boroughs of Manhattan, Bronx and Queens, but the sulfur levels in the air would probably be acceptable due to the switch from high sulfur coal and oil to desulfurated hydrocarbon fuels. Toward the end of the meeting, someone asked Mr. Kahn: What can we do about air pollution? Who are the good guys? Who are the bad guys? Mr. Kahn replied: "Oil companies; motor car companies; Con-Ed; they're the bad guys... We are the good guys."

CLEANER AIR NOW

The fight against air pollution has already resulted in the decrease of pollutants in many cities around the world. Levels of sulfur dioxide (SO_2) in New York air are lower than they were in 1965. This progress is due mainly to the banning of high-sulfur fossil fuels. Our air will have even less sulfur dioxide in it when the older power plants are replaced by atomic energy plants. It is ironic that many of the individuals who decry our polluted air are leading the fight against building nuclear-fueled plants.

The air-pollution code in New York City limits the amount of sulfur in fuel burned by power plants to 0.3 percent. Fuel oil has largely replaced coal, thus decreasing the amount of soot as well as SO_2 wafted into the air by power plants. Further reduction in non-automotive pollutants could be achieved by upgrading incinerators throughout the city and compacting rather than burning solid wastes. Compacting instead of incinerating increases the problem of solid waste disposal but construction of off-shore islands built of waste should alleviate that problem in a few years. This decrease in SO_2 levels in New York's air is typical of what is going on nationwide.

According to figures supplied by the New York City Environmental Protection Administration, particulates in the air of New York City have decreased from some 90,000 tons per year in 1965 to 36,000 tons per year in 1970, a 60 percent decline. For some still unexplained reason, suspended particulate matter rose in 1971 by 7 percent over the average of the previous two years. In 1969, the average reading taken from over 30 sampling stations in the city was 98 micrograms of suspended particulate matter per cubic meter of air; the reading rose slightly and typically to 107 in 1971. Federal air quality standards for 1975 insist on a level of 75 micrograms or less.

In January of 1973, Commissioner Fred Hart of New York's Department of Air Resources, reported that there had been an 80 percent reduction in sulfur dioxide due primarily to the upgrading of oil burners and the banning of coal as a fuel in New York City. During the past two years particulates have dropped from 107 to 74 micrograms per

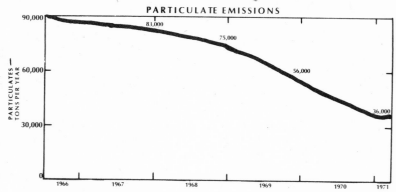

Figure 4-1   Decrease in Air Particles Over New York City
(After NYC EPA)

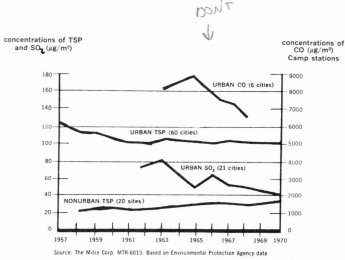

Source: The Mitre Corp. MTR-6013. Based on Environmental Protection Agency data

Figure 4-2 Trends in Ambient Levels of Selected Air Pollutants

(μg/m^3 means micrograms of pollutant per cubic meter of air)
TSP = Total Suspended Particles
SO$_2$ = Sulphur Dioxide
CO = Carbon Monoxide

cubic meter of air. The remaining pollutants in the air are caused almost entirely by motor vehicles. The clean air standards set by the federal government will probably never be met in a few downtown areas during rush hours, but I would estimate that 99 percent of New York City residents breathe air that meets the stringent Federal air quality standards over 95 percent of the time. Air quality degenerates only on four to ten lane streets during a few peak travel hours each week. Most of the time the air in New York is healthy and invigorating.

Figure 4-2, taken from Environmental Quality Council on Environmental Quality, 1972, shows quite clearly that total suspended particles (TSP) SO_2 and carbon monoxide (CO) have been decreasing steadily since 1963 in United States urban centers. Figure 4-1 shows graphically the downward trend of most urban air pollutants in recent years.

Air visibility observations give a good indication of smoke and particulates in the atmosphere. The clarity of city air in America has increased during the past few years, according to R. G. Beebe (1967), who compared visibility observations at Atlanta, Newark, Cleveland, Chicago, St. Louis, Kansas City, Omaha, and Oakland. Beebe found on the average that the number of hourly observations during which visibility was reduced by smoke or haze to below seven miles had decreased 82 percent during the 20-year period from 1945 to 1965.

In Los Angeles County, the number of days per year during which adverse levels of pollutants have been measured declined between 1956 and 1968 for each pollutant measured except oxidents of nitrogen (Lemke, et al., 1969), and even these oxides have shown a decline since 1965 in downtown Los Angeles (Air Resources Board, 1970). Between 1955 and 1968 visibility improved nearly 50 percent at Oakland Airport. Airborne particulates have declined in the Bay Area since 1957, and oxidant concentrations have declined since 1965. The U.S. Continuous Air Monitoring Program measurements in Chicago, Philadelphia, St. Louis, Washington, Cincinnati, and Denver show general downward trends in concentrations of major air pollutants since

1962 (see figure above). These are encouraging results.

The trend towards cleaner city air has been going
forward in many major cities throughout the world. For
example, in Russia, according to Soviet observer Igor
Petryanov (New York Times, August 15, 1971), "Over
the last few years, the condition of the air in many industrial
areas of the country has shown a marked improvement.
In Moscow, the air is cleaner than it was several years ago.

In capitalistic London the smog reached its peak in 1886.
Since then, frequency, density and dirtiness of smog has
declined and visibilities noticably improved, especially
since 1949 (Freeman, 1968). Less air pollution has resulted
in an increased rate of sunshine in central London.

The charts below show that smoke and SO2 have rather
steadily decreased in Paris, London, and Rotterdam during
the 1960's.

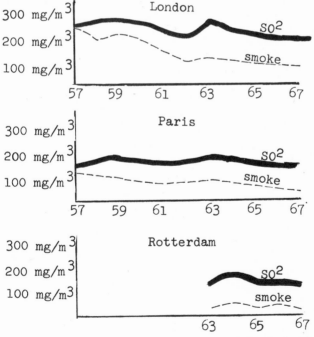

Figure 4-3   Variation in Level of SO_2 and Smoke in Three
European Cities. (Petrole Progress April 1970, 84)

Vigorous action by New York City, prodded by Citizens for Clean Air, has resulted in a marked increase in air quality in a relatively few years. Complaints against air polluters have decreased from more than 30,000 in 1969 to less than 15,000 in 1971. City air is getting cleaner in spite of dire predictions by some environmentalists and members of the air-pollution establishment.

THE DUBIOUS HEALTH HAZARDS OF AIR POLLUTION

Despite widespread attempts to link air pollutants with a variety of diseases, evidence doesn't prove that the ordinary everyday air in cities is a health hazard to average city-dwellers. Let's consider the major pollutants and their effect, if any, on health.

In 1968, Dr. Samuel Epstein testified before the Senate Subcommittee on Air and Water Pollution: "The dramatic increase in mortality from lung cancer...is now approaching epidemic proportions," as he pointed a finger at air pollution. Many people believed Dr. Epstein. But let us see what the evidence shows. Dr. Alton Ochsner, in the American Scientist (1971), commented:

"Air pollution has been blamed as the cause of respiratory disease and, although there may be a slight additive effect of air pollution, it is certainly not an important factor (in pulmonary disease). If it were, men and women should be equally affected because they both breathe the same air."

And Dr. J. R. Goldsmith (1968) wrote:

In summary, the available evidence supports a migration and an urban factor as contributing in a minor way to the complex of casual factors affecting lung cancer rates. The available evidence has failed to support the hypothesis that exposure to community air pollution is a causal factor.

Concentrations of lead in the air may vary from less than 0.01 microgram per cubic meter of air in Thule, Greenland, to 2.5 micrograms per cubic meter on some

New York City streets. Nowhere in the world is the air free of lead particles. Some of the most heavily traveled U.S. roads have concentrations of over 50 micrograms per cubic meter. But, despite seemingly high concentrations of lead in the air, most Americans inject more lead in their food and water than they do by breathing Many workers in lead industry plants breathe air containing 150 micrograms of lead per cubic meter for 40 hours a week over a period of years and have shown no ill effects over a lifetime of work.

Concerning the effects of airborne lead on health, R. A. Kehoe (1969) reported:

> It has been stated and re-stated for such a long time and in so many different media, that lead when absorbed is a "cumulative poison," as to have given this phenomena the characteristic of universality and inevitability in the popular mind, regardless of the quantities involved... In the entire literature of this subject there is not one whit of evidence that anyone has ever suffered any illness or impairment of health or well-being as a consequence of the occurrence of lead in the general human environment, as far as is known to us, admittedly somewhat incompletely, at this time in history.

The vaporous nature of much of the research on the health effects of air pollutants derives from statistical attempts to link specific sicknesses and/or mortality with specific pollutants. Statistics at best is a whorish discipline, and when small effects are sought in masses of shifting data involving dozens of variables, the skillful (or sloppy) researcher can come up with a variety of correlations, which may not be valid and bear little resemblance to the real world.

A good (or bad) example is the computer study of Alfred Hexter and John Goldsmith of the California State Department of Public Health on the association of community air pollution, especially carbon monoxide and mortality. Hexter and Goldsmith (Science, April 16, 1971) analyzed a mass of data from Los Angeles (1962-1965) and concluded that "The estimated contribution to mortality from Los Angeles County associated with CO may be a difference of as many as 11 deaths in one day, all other factors being equal." According to the two authors, death rates follow a cyclical pattern increasing in winter and decreasing in summer. But all the main smog contaminants increase during the winter because the air is colder and can hold more contaminants, and because there are more contaminants generated during the cooler months. It can be shown statistically that in Los Angeles almost all other contaminants are correlated with CO and thus also "may" cause "as many as 11 deaths in one day." Air specialists addicted to pollutants other than CO may feel slighted if death is not attributed to their particular contaminant. Since the death rate also parallels temperature changes it is also possible to show a statistical correlation between decreased temperature and increased mortality. One can also usually demonstrate statistically that increased temperatures parallel increased death rates.

It is noteworthy that the death rate in Los Angeles has been rising over the past few years, probably due to an aging population. Approximately 70,000 Angelinos die each year, so that it is easy to say, but would be extremely difficult to refute positively, the statement that in Los Angeles CO may cause "as many as 11 deaths in one day." State and federal governments in recent years have been supporting air pollution research at the rate of approximately $50 million a year, giving sustenance to thousands of Americans. It would be encouraging if the money were directed to practical problems of air pollution control rather than to attempts to justify more funds by curious statistical devices.

Consider further the statement on auto emissions of G. W. Wright (1969).

As of this moment, no data of a valid nature pertinent to the question of causation of bronchogenic cancer or diffuse obstructive pulmonary disease as related to the specific irritants being considered in this paper (oxidants, NO2 and hydrocarbons) is available... The first of these is that the lung tolerates stresses in terms of exposure to gas and particles to a degree difficult for some to believe because it has the ability to manufacture a protective coating on its surface, cleanse itself to an extraordinary degree through the mucociliary and macrophage mechanism and replace and repair its cellular constituents for many years under many conditions of stress. It is not too far fetched to believe that it was, in fact, this extraordinary ability that brought us through that evolutionary period when the environmental air may have been even more noxious than it is today.

And that old bugaboo, SO_2? Dr. M. C. Battigelli (1968) reported:

The obvious discrepancy between the alleged disastrous effect of air pollution on health and the inconspicuous concentration of sulfur dioxide measured in the air (in the Air Pollution Episodes) has taxed the imagination of toxicologists for the past 20 years.

Photochemical smog does not appear to be particularly harmful according to Tabershaw, et al. (1968):

The fact that millions of Los Angeleans have been exposed for years to above-threshold limits of oxidant without any serious health effects raises doubt that we are correctly assessing the toxicity of this pollutant in humans as it exists in smog.

Among the conclusions on health hazards reached by the prestigious California Air Resources Board, Department of Public Health (November, 1970) are the following:

No short-term increase in general mortality from photochemical pollution has been found or is expected.
Although there is suggestive evidence that carbon monoxide pollution may lead to increased fatalities among persons having heart attacks, information available at this time is not adequate to evaluate the need

for emergency action by State or local government...

"There is no evidence that respiratory cancer is associated with community air pollution in California." And since Los Angeles has the worst imaginable meteorological conditions it is unlikely that other communities will be harmed.

With regard to other lung conditions, all the Board would venture is that "Air pollution in California <u>may</u> <u>have a role</u> in causing or significantly aggravating emphysema and other non-malignant respiratory diseases." This, after millions of dollars and thousands of man-years have been spent in air pollution research.

Dr. Hugh Elsaesser (1971) has made an extensive study of the scientific literature on the supposedly injurious effects of air pollution to humans. He wrote:

> Despite many recent claims to the contrary, currently available evidence fails to confirm the contention that we are in imminent danger of rendering our planet unfit for life because of air pollution...
>
> If one examines the thousands of experimental and epidemiological studies published in search of health effects of air pollution, he finds it difficult to maintain that the above statements are due to lack of research. One should also note that researchers finding negative results (no health effects) experience far more difficulty in securing contract renewal and publication than do those who claim positive results. Inability to detect a health effect has usually been interpreted as a project failure rather than as evidence against the existence of such effects. Thus, such results are rarely disseminated beyond a minimum edition contract report and a conference presentation...

While the average citizen can breathe city air with impunity, certain types of foul air combined with peculiar meteorological conditions (called episodes) have led to increased mortality rates. Two dramatic episodes that demonstrated an increase in diseases of the lungs occurred in Donora, Pennsylvania, in October 1948 and in London in 1952. In both cases, the air, containing pollutants from burning coal, stagnated above the cities for more than three days. The increase in deaths in Donora and London occurred almost exclusively among older persons with previous bronchopulmonary diseases. There is little question that stagnant air heavily contaminated with sulfurous and nitrous oxides or carbon monoxide can be injurious. I have heard, but have not been able to verify, the assertion that the increased death rates during the Donora and London episodes were followed by decreased death rates, which might indicate that individuals on their way to Hades had their trip speeded up by the episodes. But, from the studies cited here, it would appear that everyday city air does not ordinarily contain a lethal quantity of poisons.

As die we all must, 1,243 men and 388 women died, and Dr. Oscar Auerbach and his associates at the Veteran's Administration Hospital in East Orange, New Jersey, (1972) dissected the corpses.

Pieces of the lungs of the dead men and women were cut out and the sections were studied under microscopes for indications of lung disease. The researchers found no noticable difference between the lungs of the persons who had lived in the "dirty" city and those of persons who had lived in the "clean" country, indicating there is no increase in the incidence of lung cancer or emphezema due to breathing city air.

In a letter to me dated June 14, 1972, Dr. Auerbach wrote: "We have found that if one smokes cigarettes in the city or in the country, there is no difference in the changes within the tracheo-bronchial tree or in the lung parenchyma." And further: "I would like to specifically state that we found lung cancer was in no way associated with city dwelling if the man did not smoke."

SOME AIRY REMARKS

City dwellers who, after reading this, are still disturbed by the carbon monoxide and sulfur dioxide in their air (and are willing to change their modus operandi), may take the subways instead of cars or buses. The air in the subway tunnels may leave much to be desired, but it is likely to have lower concentrations of some of our more undesirable pollutants that are generated at and above street level. In addition, the ozone generated by the third rail kills pathogens and helps sweeten the subterranean atmosphere.

The average Manhattanite lives and works in high-rise buildings and spends much of his life 50 feet above the ground, where the air is purer than it is at street level. An increasing number of city dwellers spend only a few hours a day at ground level in the open air. This rise into the atmosphere is probably good for their health.

New York City's Department of Air Resources (DAR) issues a daily bulletin rating city air on a scale of "good," "acceptable," and "unhealthy." According to the DAR there were no "good" days in New York City during 1970 and 60 days were labeled as "unhealthy." During 1971 the DAR assures us that there were some days with "good" quality air. According to air alarmists, there should be a correlation between death rates and air so that fewer people should have died in 1971 than in 1970. This has not been shown.

City dwellers, particularly those in New York, who are fearful that their air is being poisoned, should note that the standards set by the city for describing air as "unhealthy" are often quite arbitrary and in many cases misleading. For example, designations are based on samples of carbon monoxide taken from some of the worst possible locations, on streets with high traffic densities. If the samples were taken two blocks away on an uncrowded side street, the reading would most likely be "good." The pollution scale is based on city-wide averages of air, sulfur dioxide, soot, carbon monoxide, and photochemical oxidants, a major ingredient of smog. Air termed "unhealthy" means only that breathing it for unspecified long periods may sicken some susceptible individuals. But city officials concede that it's practically impossible to pin down just what "long periods" might mean, and as we have shown, the supposed injurious effects on the average citizen of breathing city air are unproven or illusory.

Actually, some pollution may be beneficial to health. About 120,000 Americans will develop skin cancer in 1971 primarily from exposure to the sun's ultraviolet rays. A recent study by the Smithsonian Institution, as

'You'll be all right in a few days. Just stay indoors, keep the windows closed, and stop reading the New York Times Air Pollution Index.'

The New Yorker

reported by S. Dillon Ripley in the American Scientist (September-October 1971), showed that ultraviolet radiation in downtown Washington has decreased by about 16 percent since 1907. Air pollutants are salubrious to the extent that they temper the sun's shorter wave-lengths and shield light-skinned individuals susceptible to skin cancer. There are probably thousands of Washingtonians and Angelinos who owe their clear skins and good health to the fact that they have been protected from direct sunlight by their city's smoggy atmosphere.

Polluted air is hardly likely to be beneficial, but that is not the same as saying it is injurious. Dr. Ochsner (1971) noted, "Air pollution is undesirable, and under certain adverse meterological conditions may be a hazard to certain individuals with severe cardiopulmonary disease, but generally it is not a great health hazard because of the relatively low concentration of pollutant as compared with that in personal air pollution from cigarette smoking." This conclusion concurs with that of the National Advisory Council on Cancer.

Perhaps the lack of disease due to air pollutants derives from the ability of the human body to ignore, or at least naturally recover from, minor insults. Small doses of poisons such as carbon monoxide appear to have little effect on the human body, whereas breathing CO in concentrations greater than 20 parts per million can make one nauseous; and much greater concentrations will lay one out for good. City dwellers breathe only relatively small concentrations of pollutants and their effect on health appears to be minimal. As noted before, these concentrations are decreasing in most large cities.

5

THE MYTH OF VANISHING OXYGEN .

The myth of vanishing oxygen goes like this: Pesticides, specifically DDT, are reducing the phytoplankton population in the ocean and thus reducing our planet's oxygen supply (C. F. Wurster, 1968). In reality, only about 1/20 of one percent of the earth's atmosphere is renewed each year by photosynthesis, and only half of that comes from the ocean.

Another exciting theory allows that man's burning of fossil fuels is using up our planet's oxygen. Or to quote A. Toffler in <u>Future Shock</u>:

> On land we concentrate such large masses of population in such small urban technological islands, that we threaten to use up the air's oxygen faster than it can be replaced, conjuring up the possibility of new Saharas where the cities are now.

Since oxygen is an essential element for life, one would have cause for alarm, if these predictions were valid.

OXYGEN IN OUR ENVIRONMENT

Our planet is surrounded by a layer of compressible gasses, primarily nitrogen (80 percent) and oxygen (20 percent). Other gasses regularly found in relatively small, but variable quantities in the earth's atmosphere are water vapor, methane, carbon dioxide, carbon monoxide, helium, and ozone. Industrial processes have added trace gasses, such as sulfur dioxide, oxides of nitrogen, and carbon monoxide, and particulates such as lead, soot, and asbestos.

Oxygen is generated by green plants in the presence of light. In the basic photosynthetic process carbon dioxide and water are turned into oxygen and plant material: CO_2 + H_2O---Oxygen plus carbohydrates usually in the form of sugar or starch.

Algae in the ocean as well as plants on land generate oxygen during daylight hours. Although the ocean covers 71 percent of the earth's surface, less than 50 percent of the world's oxygen is generated at sea. Oxygen is consumed or converted into CO_2 or CO by burning, respiration, and bacterial decomposition of organic matter. In nature the release of oxygen is approximately balanced by its consumption on a worldwide scale. But some areas such as coastal waters and forests generate more oxygen than cities, which consume more than they generate. Decaying organic matter removes oxygen from a water body. However, healthy algae growth adds oxygen to water.

LAYING THE MYTH TO REST

C. F. Wurster's laboratory study on the effects of DDT on phytoplankton reported that one part per billion (1-ppb) of DDT had no effect on four species of marine phytoplankton. He found that 10 ppb interfered with plankton photosynthesis and growth. What Wurster did not say is that his study is irrelevant to the real world since the levels of DDT plus its metabolized forms in ocean water are only one part per trillion or less. In all fairness, Wurster might have noted that his results indicated that DDT is not likely to effect phytoplankton production now or in the future.

In a talk given at Yale (1970), Wurster said of his laboratory studies of the effect of DDT on algae, "The data indicated that as DDT was added to water, the rate of photosynthesis was decreased. Photosynthesis is the process whereby green plants absorb carbon dioxide and the energy from sunlight, producing organic nutrients and oxygen. All animal life on earth is dependent on this process." Wurster may never have believed that DDT could diminish the earth's oxygen supply, but his juxtaposition of sentences leaves the impression that this is the case. It is unlikely that a host of scientists would have devoted their efforts to disposing of this fantasy if Wurster and his rabid allies had not promoted it.

Paul Ehrlich elaborated and disseminated the myth for greater public consumption in the September 1969 lead article in Ramparts, entitled "The End Of The Ocean." Ehrlich wrote:

> The end of the Ocean came late in the summer of 1979, and it came even more rapidly than the biologists had expected. There had been signs for more than a decade, commencing with the discovery in 1968 that DDT slows down photosynthesis in marine plant life. It was announced in a short paper in the technical journal, Science, but to ecologists it smacked of doomsday. They knew that all life in the sea depends on photosynthesis, the chemical process by which green plants bind the sun's energy and make it available to living things. And they knew that DDT and similar chlorinated hydrocarbons had polluted the entire...

The fantasy was picked up and widely publicized by U Thant of the United Nations and other frightened and misinformed people.

A comprehensive study of the availability of oxygen in our atmosphere was performed by Lester Machta of the National Oceanic and Atmospheric Administration (NOAA), and Ernest Hughes of the National Bureau of Standards (1970). Machta and Hughes reviewed oxygen measurements in 88 samples of air collected at sea and over land between 1967 and 1970. They found no decrease in the element vital to life. With only one exception, the percentage of oxygen remained over 20.94. They determined that the level of oxygen in the atmosphere has remained constant for at least the past 60 years; 20.94 percent of our air by volume consists of oxygen molecules and this percentage has not changed.

In fact, Machta and Hughes have calculated that if all the recoverable fossil fuel (oil, natural gas, and coal) in the world were to be burned, the combustion would reduce the oxygen level to 20.80 percent, an insignificant change. In fact, persons living at an altitude 250 feet above sea level breathe air with a partial pressure of oxygen of 20.8 percent.

Another article, "Man's Oxygen Reserves," by Wallace Broecker (1970), helped to dispel the unrealistic fear of vanishing oxygen. Dr. Broecker, of Columbia's Lamont-Doherty Geological Observatory, used geophysical and geochemical data to show that the oxygen myth is pure illusion. Broecker showed that even in large urban centers oxygen depletion is a minor problem. In fact, his calculations indicate that city residents would be poisoned by carbon monoxide from auto exhausts long before the oxygen content of the air had dropped by as much as two percent.

If photosynthesis were to cease entirely, less than one percent of atmospheric oxygen would be used up to oxidize all organic debris. If all of the sewage wastes of two billion humans were dumped into the sea, the oxygen in sea water right now could assimilate this waste for over 12,000 years.

Man's greening by irrigation of desert areas throughout the arid regions of the world has added to the supply of oxygen in the atmosphere. Proper fertilizing of oceans and lakes with sewage causes increased algae blooms, which tend to further increase the world's oxygen supply.

The relative quiescence of the adherents of ecological doom on the subject of atmospheric oxygen depletion would seem to indicate that some environmental myths can be laid to rest by scientific reasonableness. If one must be anxious about deterioration of the environment (and there is good cause to be for some pollutants in some areas), the concern should be with problems other than oxygen.

THE "DEATH" OF LAKE ERIE

The beginning of wisdom is to call things
by their right names.
                                    Chinese Proverb

 Thirty booby birds
 went skimming o'er the wave.
 We chased them and we chased them
 till our victuals gave.'
                                    Peter Agnos

  The ecological myth of Lake Erie in its starkest form
says that Lake Erie is a dead body of water. According to
René Dubos, "Lake Erie has been turned into a cesspool "
Barry Commoner uses the myth of the dead lake to warn
us of the gruesome end to the earth as a place to live:
"Lake Erie represents the first large-scale warning that
we are in danger of destroying the habitability of the earth.
Mankind is in an environmental crises and Lake Erie
constitutes the biggest warning." In Science and Survival
(1963) Dr. Commoner, called the "Paul Revere of the
environmental movement" by some of his admirers, wrote
(p. 12), "...Lake Erie has already been overwhelmed by
pollutants and has, in effect, died... The fish are all but
gone..."
  The press's attitude towards Lake Erie (and many other
environmental problems) is typified by the color film,
"Who Killed Lake Erie?," which is described by NBC
Educational Enterprises as "a comprehensive introduction
to current environmental disasters, concentrating on the
murder of Lake Erie by people from all walks of life."
  What does the designation "dead lake" mean to the
average citizen? I put this question to students in my class
at the New School during our first meeting in February of
1972. They said it means "there is no life in the lake;"

that "the lake is so polluted that it cannot be used for anything;" that "all the fish are dead;" that "algae are killing
all the life in the lake." I would venture to say the Average
American would agree with most of these mournful designations as applying to Lake Erie

Lake Erie has a surface area of approximately 10,000
square miles. Its maximum depth is 210 feet with an
average depth of 58 feet. There are 120 trillion gallons
(110 cubic miles) of water in the lake. The Detroit River
with a mean flow of 125,000 million gallons per day (190,000
cubic feet per second) flows into the lake. The lake drains
near Buffalo via the Niagara River over Niagara Falls into
Lake Ontario. (See map)

The lake is divided naturally into three main basins.
The western basin covers some 1,200 square miles. It is
extremely shallow with an average depth of only 24 feet.
The water is turbid and there is an enormous productivity
of plant life. The central basin covers approximately 6,300
square miles and has an average depth of 60 feet with good,
clear water. The eastern basin consists of the remaining
2,400 square miles. It contains the deepest and coldest

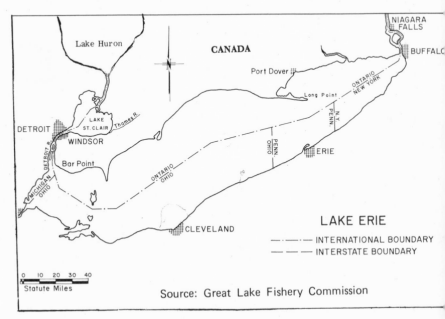

LAKE ERIE

—·—·— INTERNATIONAL BOUNDARY
— — — INTERSTATE BOUNDARY

Source: Great Lake Fishery Commission

lake water.   Here is found the lake's greatest depth, 216 feet.   The water of the lake is generally good to drink and swim in except for a relatively few miles fronting along the south shore in the vicinity of presently untreated sewage outflows.

On and around the lake, tourism is a major industry which presently adds hundreds of millions of dollars annually to the economy of the Erie basin.   Millions of citizens use lake beaches every summer, fish in the productive waters, boat or simply walk along the shore.   Iceskating and iceboat racing are the big attractions during the winter. On a warm summer weekend some 20,000 pleasure boats can usually be found on the waters of Lake Erie (FWPCA 1968).

Lake Erie waters are used for municipal and industrial, as well as agricultural water supplies, cooling of power plants, recreation, navigation, commercial fisheries and wild life.   In addition, the lake receives domestic and industrial waste waters, and the runoff from agricultural lands.

FRESH WATER USE

Lake Erie is the prime source of drinking water for Cleveland, Lorain and Toledo, Ohio; Erie, Pennsylvania; and Buffalo, New York.   Approximately three and a half million people use an average of 650 million gallons of Lake Erie water a day.   On peak days, water use exceeds a billion gallons.   No health hazards have developed, or are foreseen by sanitary engineers, for continued use by millions of people of Lake Erie water.

The International Joint Commission of Canada and the United States in its 1970 report on Lake Erie notes that: "The waters of Lake Erie two or more miles offshore can be classified as bacteriologically excellent...and most bacterial pollution appears to be localized."

Widespread use of Lake Erie water by man and animals is hardly indicative of toxic or grossly polluted waters. It is nevertheless true that along the south shore of Lake Erie generally, and in the particular vicinity of the industrial center of Cleveland, and near some of the other industrial areas, run near-shore currents of often wretched

113

smelling and evil-looking water. However, in perspective, the vast majority of lake water is clean and potable. And even Cleveland, with this local, near-shore problem, is able to use more than 400 million gallons of lake waters per day, requiring no more than average water supply treatment.

Cleveland, the major city on the lake's south shore, draws all of its drinking water directly from the lake at a point approximately three miles from shore. Sanitary engineers who monitor the water have found that the water drawn from the lake is almost completely free of coliform bacteria throughout most of the year. Coliforms are used as indicators of animal and human fecal pollution. A few times during the summer the coliform count may rise probably due to boats discharging waste in the area of the intake pipe. But the waters of Lake Erie are expected to remain clean and potable for the foreseeable future, and Cleveland plans to draw all its future water supplies from the lake.

TABLE 6-1
Commercial Fish Catch from Lake Erie (1968)
Figures supplied by Great Lakes Fishery Commission, 1970
(in thousands of pounds)

	U.S.	Canada
smelt	1	12,223
carp	2,683	94
white bass	728	750
yellow perch *	3,735	24,435
sheepshead	3,144	651
All species	11,920	39,415

* highest catch on record.

FISH

Fish are excellent indicators of pollution since many species cannot survive in toxic waters or waters devoid or overpopulated by algae and/or waters with very low

114

oxygen levels. Microscopic phytoplankton (drifting uni-cellular plants) form the base of the aquatic food chain. Zooplankton (microscopic animals) and small fish eat the phytoplankton, and these in turn are eaten by larger fish. Only when a body of water provides the proper food, physical environment, and oxygen will fish thrive. These conditions exist in Lake Erie as proved by the extensive fish population now living there.

Lake Erie abounds with smelt, yellow perch, shad and white perch (Table 6-1) though these are not the fish that dominated the lake 30 years ago. One also finds smallmouth bass, carp, sunfish, walleye, and northern pike in profusion. The lamprey, poor American fishery management and an increase in nutrients have caused the whitefish, walleye, cisco, and blue pike to be replaced by other species. Sauger and lake trout have declined, but overall there are now more fish in Lake Erie than there were 30 years ago. Lake Erie fish are edible and rela-tively free from pesticide and heavy metal contamination.

The annual catch in Lake Erie during the 1930's averaged 42 million pounds. What has been the effect of the "pollution" that has turned Erie into "a dead lake?" According to Carl Parker of the New York State Depart-ment of Environmental Conservation (letter November 9, 1971), the total fish catch taken from Lake Erie during recent years has averaged over 100 million pounds. "The commercial catch has averaged over 50 million pounds for many years. The sport catch is probably much larger. The major change in the past 20-30 years has been the decline in coldwater species which used to inhabit the depths of the lake..." Commercial fishing reached a peak in 1969 of about 59 million pounds The decline to 42 million pounds in 1970 was in large part due to fear induced in consumers by the mercury poisoning scare. A black market has grown up in parts of Ohio for certain Lake Erie fish over the size allowed by the State conservation authorities tuned into the mercury-in-fish chorus. About three-quarters of the commercial catch is taken by Cana-dian fishermen who have the advantage of government support, better gear and less interference from environ-mentalists compared to their American neighbors.

Lake Erie supports a population of fish larger than all the other Great Lakes combined! By contrast, Lake Superior with its sparkling clear water supports virtually no fish. Nutrients and organic matter from runoff of surrounding terrain are trapped in the shallow western portion of Lake Erie and here algae flourish, thus providing the needed food for zooplankton and ultimately all marine life in the lake. Fish thrive on the green plant life in the water. Some bathers may be annoyed by the organically rich, greenish waters, but it is no more justifiable to say algae-filled waters are a desert than to say a rich field of grass is a desert.

POLLUTANTS IN LAKE ERIE

To show that Lake Erie is alive and kicking is not to gainsay the fact that there are unwanted materials, i.e., pollutants, in the lake. Prime among the pollutants are waste petroleum products. Relatively large amounts of waste oil are spilled into Lake Erie from the industrialized communities along its shores. For example, in 1969, the Detroit River dumped an estimated 1,000 barrels per day of oil into the lake. The Buffalo River is another prime source of oil, as was Cleveland Harbor. The small river flowing past Cleveland occasionally caught fire due to inflammables on its surface. An attempt is being made to control Cleveland's waste floatables by a diked disposal area.

Three metropolitan industrial areas are primarily responsible for whatever poor conditions exist in the lake. The major contributor of phosphorus, Biochemical Oxygen Demand (BOD) and chlorides is the Detroit and Southeast Michigan industrial region which contributes 40 percent of the phosphorus, 60 percent of the BOD and 51 percent of the chlorides. The Cleveland-Akron area and the Toledo-Maumee River area are the other two major sources of these particular pollutants.

Municipal sewage is the prime source of Lake Erie nutrients, such as phosphorus (which may help trigger algae blooms), although agricultural runoff in the western part of the lake is also large. According to the International Joint Commission, 70 percent of the phosphorus

116

in U.S. sewage comes from detergents and about 25 percent from human waste; while in Canada 50 percent of the phosphorus derives from detergents and 50 percent from human waste. Phosphorus originates from agricultural runoffs and industrial wastes for a total phosphorus load in 1967 of 60 million pounds. This situation is improving as new waste water treatments plants come into use.

EUTROPHICATION

The term "eutrophic" meaning well-nourished, was introduced into the scientific literature by C. A. Weber in 1907 to help explain some German bogs he was studying. Einar Neumann introduced the term into the study of limnology (the science of lakes) in 1919 to indicate lakes with high nutrient concentrations. Well-nourished lakes with abundent amounts of nitrogen, phosphorus and other plant nutrients, usually have high rates of plant growth and algae and the fish that feed on them. The phrase "cultural eutrophication" is used to distinguish enrichment of a water body due to man's activities, rather than enrichment caused by natural processes. But in any case, eutrophication indicates an aging process that most, but not all lakes pass through on their way to becoming bogs, plains and forests.

In nature, environmental change in lakes occurs inevitably through fertilization. This phenomenon, basically a natural process, is often accelerated by human activity. Most lakes at the time of their formation were probably oligotrophic, that is, almost free of inorganic nutrients and practically all dissolved organics. Only limited biological production can take place in an oligotrophic lake. Living matter dies, and then decomposes. The nutrients become available slowly, and the nutrient value of the water increases.

In the course of time, usually over thousands of years, the life in the lake gradually changes: Nutrients concentrate in the water, due to processes in the lake and runoff from the land. The lake slowly undergoes a process called eutrophication. The type of plants and fishes in the lake changes. Nutrient enrichment gives rise to higher levels of biological activity. Eventually, the lake's activity may

117

cause accumulation of large quantities of organic matter that cannot readily be decomposed by oxygen in the water. Then particles sink to the lake bottom and anaerobic (oxygen free) sediments accumulate. The sediments build up to the point where swamps or marsh lands develop. Erosion of land around the lake often accelerates the filling up of the lake. One result of eutrophication is that some so-called desirable cold-water species of fish are replaced by forms not fully utilized by American sportsmen. Man's activity in dumping sewage, spoil and nutrients into a lake speeds up the inevitable, natural process of eutrophication.

The man-made changes of Lake Erie represent only a slight shift in the natural balance of the environment. Sewage and chemicals from the cities and industrial centers around the lake have speeded up eutrophic conditions by perhaps 200 years. At any rate Niagara Falls will erode back in a few thousand years converting Lake Erie into a wide river. In one sense, man has helped nature along the road she was travelling. The lake trout has been replaced by white perch bass and pike as the productivity of the lake has increased and the lake now supports a fish population considerably greater than that of 30 years ago. There is no reason why existing fish cannot be harvested for human consumption. Indeed, the Canadians are doing just that.

THE FUTURE OF LAKE ERIE

The furor raised over the "Lake Erie is dead" myth has led to much study and attempts to clean up oil spills and industrial and municipal pollution in the Great Lakes. In June, 1971, Canada and the U.S. announced agreement on a joint program to end water pollution in the Great Lakes and the St. Lawrence River by 1975 -- to make those waters "clean enough for any fish to live in," a rather flippant remark, in light of the happy fish swimming in Lake Erie now.

The clean-up program, which will cost between $2 and $3 billion, calls for the construction of treatment facilities for municipal and industrial waste, reduction of phosphorus discharges, elimination of mercury and other toxic-heavy metals from discharges, control of

118

thermal, radioactive and pesticide pollution." (The New York Times, January 13, 1971)

One can certainly agree with these efforts even though they were to a great extent fostered by ecological misstatements.

But despite some pollution, it should be obvious, even to the casual reader, that Lake Erie with its large fish populations, its low bacterial count, and enormous quantities of clean potable water, is far from dead. Millions of Americans living near the lake swim, fish and boat in Erie's waters each year. Sport fishing alone is a multi-million dollar business. Paul Ehrlich, in his widely read book, The Population Bomb, wrote:

> You see, Lake Erie has died. The lake can no longer support organisms that require clean, oxygen-rich water. Much of this shallow body of water is a stinking mess, more reminiscent of a septic tank than the beautiful lake it once was..." No one in his right mind would eat a Lake Erie fish today.

One wonders who is "in his right mind," and what forces drive persons such as Ehrlich to publish such nonsense.

There is a way to slow down, but not stop, the natural filling in of Lake Erie by diking the shallow western end of the lake. Advanced diking techniques have been developed by the Dutch in their long struggle with the North Sea. Once the dikes are built, waste sludges, sediments and waste effluents could then be piped to the diked area that would act as a settling lagoon. The pond would first turn to a swamp, to the delight of water birds. In the natural course of events the swamp becomes a bog And in time the bog turns into a meadow. Eventually some trees take root and a forest grows. What we have here prescribed for Lake Erie would simply and naturally result in a progression of ecological regimes. In the process, we would create for our great grandchildren one of the largest national parks east of the Mississippi. In the long run, diking and piping probably would cost less than building hundreds of sewage treatment plants around the Lake, plants that do not remove all nutrients and create a sludge disposal problem, that still remains to be solved.

7

OIL ON THE SEA

The oil spill myth goes something like this: The ocean is rapidly becoming irreparably contaminated by oil spills. This is exemplified by Jacques Cousteau's statement that 40 percent of all life in the sea has been eliminated by man's activities. Max Blumer (1971) stated more cautiously that "all crude oils and all oil fractions except highly purified and pure materials are poisonous to all marine organisms."

Typical of the dire predictions that followed the Santa Barbara oil blowout were comments in the <u>Santa Barbara News Press</u>: "Pollution of the offshore and inshore waters could kill off plankton and other marine life..." A local columnist, Henry Ewald, wrote, "sport fishing and commercial fishing will be ruined for years." And oil contamination "...will keep out the migratory fish for years." A California marine biologist predicted that the oil would kill all marine life along 20 miles of coast.

The oil myth is widely repeated by the press and believed by the public because black oil on water makes a dramatic photograph. Oil looks and smells bad, and the fact that the oil may not harm life in the water below is not apparent to the uninformed TV viewer. Oil slicks are misleading in that a water body's surface appearance often has little to do with water quality. A lake that appears crystal clear may be poisonous to fish and man, while water covered with oil, debris, and litter may be potable and support a healthy fish population. In any case there remains the myth of ocean pollution due to oil spills.

WHERE THE OIL COMES FROM

Approximately two to ten* million tons of oil found its way into the ocean during 1971 through ocean shipping, offshore drilling, natural seepage, accidents, and runoff

* *Estimates differ greatly, indicating the vagaries of the spill problem.*

of waste oil from automobiles. An unknown but probably larger quantity of hydrocarbons reached the ocean through atmospheric fallout. There are six major sources of direct nonatmospheric oil-on-water pollution.

1) Catastrophic tanker accidents: The <u>Torrey Canyon</u> spilled 117,000 tons of Kuwait crude oil when it ran aground off the southwest coast of England in March, 1967. It soon became clear that the spill presented a grave threat to the bathing beaches and a grim warning of the dangers of the new era of super tankers. Dutch salvers attempted unsuccessfully to refloat the ship. The Royal Navy tried to confine the oil to the area of the broken grounded ship. All efforts proved unsuccessful. Finally, British planes bombed the ship and set fire to most of the remaining oil. By then, the sticky oil had spread over the beaches and rocky inlets of Cornwall. Massive doses of detergents were sprayed on the oil-soaked beaches and nearby waters to emulsify the oil. These detergents did more damage than the oil to fish, mollusks and marine plants. Great Britain spent more than $6 million in clean-up operations. In April, oil reached the coast of Brittany. The French used chalk in attempts to sink it, but the damage to marine life was still considerable

By the summer of 1968, the sea had washed the British and French beaches clean and most marine life reappeared. Seaweed-eating shellfish were relatively scarce in some coves, resulting in seabeds covered with unusually thick carpets of seaweed. By 1971, the area had returned to normal.

The <u>Torrey Canyon</u> was only about a half to a third the size of today's supertankers. About one billion metric tons or 60 percent of the world's yearly oil production is transported by ocean-going vessels. During the 1960's, there were over 550 tanker collisions in American waters, four-fifths of which involved ships entering or leaving ports.

Of the 1,820 million metric tons of oil produced in 1969, 1,180 million tons crude and products were transported in tankers. Oceanborne oil is expected to rise to 1,820 million tons by 1975, and 2,700 million tons by

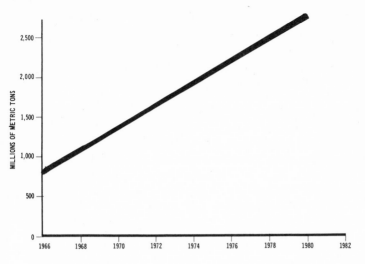

Figure 7-1 Oil Transported at Sea Annually & Projections to 1980.

1980. 500,000 tons of all the oil spilled on the sea during 1970 came from tanker accidents.

2) Offshore oil drilling accidents: The Union Oil Company's drilling rig accident off the Santa Barbara coast in 1969 is a good example of such pollution. Before the leak could be plugged, 10,000 to 12,000 tons of oil escaped to form a drifting ocean swath 10 miles long and 200 feet wide. Some of the oil reached the beaches. The seepage was reduced as the year went on. Brigades of oil company workers with bulldozers and rakes cleared the beaches. The spillage continued at a rate estimated between 30 and 100 barrels a day. Incidentally, natural seepage in the area has been estimated at 100 barrels per day (Allen et al, 1970). Union Oil claims to have spent $4 5 million in six months cleaning up more than 18 miles of beaches. The State of California and the City of Santa Barbara are suing Union Oil for over $1 billion in damages.

3) Deliberate cleaning of tanks and dumping of oily wastes at sea by ship operators accounts for about a half-million tons of oil a year. This spillage is pumped from the bilges and tanks of the world's 4,000 tankers and 30,000 other large ships sailing the world's oceans. Voluntary "load-on-top" (LOT) practice in the past several

years has resulted in reducing oily waste dumped in the sea from an estimated 1,000,000 tons annually to 40,000 tons annually. It is not usually possible to determine the source of most of the oil spilled in this fashion. At the 1971 annual meeting of the American Chemical Society in Los Angeles, various methods were proposed for tagging the oil as it moves across the ocean in ships, but as of now, there is no sure way to place the blame.

4) Chronic spillage in the vicinity of shore-based oil-tank farms and refineries: There are about 100 oil-handling terminals, 100 commercial and naval shipyards and over 60 oil tank cleaning firms in the U.S., where oil in the past was liberally disposed of into the sea during cleaning operations and from accidental spills. Stricter laws and use of oil-water separators have reduced this particular type of spill.

5) Industrial and automotive waste oil: A major contribution to oceanic oil pollution are the waste lubricants from industry and from engine vehicles. About 43 percent of the petroleum products sold in the U.S. are used for automobiles. Approximately 1.2 billion gallons of waste oil from cars and industry are disposed of each year by dumping, burning, or reprocessing. Economics now dictate that most waste oil is not reused but thrown away. Much of this waste oil presumably winds up in the sea. Glazier and Somner (1970) estimate that between 500,000 and one million metric tons of automotive waste oils were disposed of in 1969. About one million tons a year of waste oils are generated by industry. Much of this reaches the sea by river and sewer runoff.

6) Another major source of oceanic oil pollution are the hydrocarbons given off in the exhausts of internal combustion engines. It is estimated that 90 million tons of hydrocarbons from petroleum origins are put into the air each year. No one really knows, but if only 10 percent of these hydrocarbons reach the sea, they would contribute four times as much oil to the ocean as all other sources combined.

SOURCES OF HYDROCARBONS FLOWING INTO
THE OCEAN
(1970 Estimate)

	Metric tons/year
Tankers	550,000
Other ships	500,000
Offshore production	100,000
Accidents–Ships	100,000
Nonships	100,000
Refineries	300,000
Run-off from land	450,000 *
Subtotal	2,100,000
Airborne hydrocarbons	9,000,000
TOTAL	11,100,000

Source: Man's Impact on the Global Environment

According to Porricelli et al. (1971), 4.9 million tons
of oil were spilled into the world's oceans by tankers,
accounting for 28.4 percent. Tank barges account for
another 1.4 percent, and other vessels for 17.3 percent.
But the largest contribution of spilled oil is not from ships
at all. Waste automobile crankcase oil accounts for 29,4
percent of all spilled oil, industrial machinery waste oil
for 15 3 percent and refinery spills for only 6.1 percent.
Offshore drilling adds only 2.1 percent to the total spillage.

The most dramatic oil spills of the Santa Barbara and
Torrey-Canyon types which generate the most press coverage are responsible for only a minor fraction of the
total hydrocarbons entering the world's oceans.

EFFECTS OF OIL ON MARINE LIFE
Humans use petroleum products intimately, putting
pomades on their lips to prevent chapping, applying balms
on wounds to promote healing, and drinking mineral oil to
aid digestion. Thus, it would appear that not all oil
products are harmful per se to man and some in fact
promote personal health. However, some fractions of

* Porricelli et al., 1971 estimate this at 1,400,000 tons

crude oil are toxic to certain marine organisms under certain conditions. Two recent reports by Blumer and by Straughan point up the conditions that can lead to marine kills, and where and when oil spills are not likely to cause damage.

Most oil spills occur near land in estuaries and coastal waters—areas of prime importance to plant life, shellfish, shore birds, and fish. Oil contamination of the seas represents a particular hazard to many species of sea birds. Waterfowl are attracted to oil slicks, which appear to be calm water, and often dive into their midst. Birds cannot fly with oily wings and can neither forage nor escape from predators. Oiled birds have a low probability of survival. The decline of the British auk over the past 30 years is usually attributed to oil spills. British scientists have estimated that many waterfowl, 20,000 guillemots and 5,000 zargells died as a result of the Torrey Canyon disaster alone. Birds have relatively high body temperature and their bodies are insulated from cold air and water by air pockets in their plumage. A one inch spot of oil in a bird's feathers may make it impossible for the bird to maintain its normal temperature and make the bird susceptible to infection and early death.

From an ecological point of view, the use of detergents to disperse the oil makes the oil more available to the marine biota, and most of the detergents that have been used are more toxic than the oil itself. Sinking the oil to the bottom moves the problem to the ocean floor where toxicity may result in as deleterious effects to bottom-dwelling organisms.

A number of extensive studies have been conducted investigating the biological effects of oil spills in laboratories or in natural condition. From 1947 to 1950, J. G. Macklin at Texas A & M University investigated the effect of oil spills on aquatic vegetation, including those on the shore. Macklin submerged aquatic plants in crude oil in their natural environment and then determined oil levels in the mud after exposure. He found that no serious damage could be attributed to the oil, either on the vegetation or on the associated fauna. Similar conclusions were reached in extensive studies by British investigators after the spill

126

from the Torrey Canyon. Sensitive planktonic organisms seem to be affected by crude spills and could suffer or die off. The extensive damage that occurred on the shores could be ascribed mainly to the use of large quantities of detergents.

After the Santa Barbara oil leak, most biological studies indicated that the crude oil washed ashore affected the existing fauna and flora only slightly if at all (Jones et al., 1969). By the time the oil reached the shore, toxic volatile components were lost and the slight biological damage that did occur was mostly due to smothering action of the oil. Both in the Torrey Canyon and in the Santa Barbara spills, local bird fauna suffered extensively. The elimination of several thousand birds in a locality may bring about a serious, though not permanent, change in the food relationship of local organisms. For instance, the death of many sea birds may upset an ecosystem by allowing the rapid population increase of certain shellfish and finfish. Algal vegetation on exposed rocks may be smothered and eliminated by oil. Upon the death of these organisms, hardier and faster growing forms, such as sea lettuce, may completely take over an area and result in a one-sided population which may or may not be bad.

One of the most extensive studies on the effects of oil spills was conducted by Dr. Dale Straughan of the University of Southern California's Allan Hancock Foundation, in a 900-page, $250,000, year-long, two volume study financed chiefly by the Western Oil and Gas Association. The study has been widely cited as evidence that offshore oil exploration is not likely to cause permanent damage to the ecology.

Dr. Straughan reported that it is often difficult to isolate effects of oil pollution from other phenomena. Damage to flora and fauna in the Santa Barbara Channel was much less than predicted, and the area recovered well. Mortality in the channel from direct toxic effects of the spilled oil was low. The only non-avion marine species seriously affected by the spill was Chthalmalus fissus, a barnacle. Sea birds had high mortality rates, probably because of their high level of contact with the oil. Overall fishery landings at Santa Barbara in 1969 were similar to those of 1968, indicating that fish populations were probably not affected by the spill. Straughan examined various theories to explain why so little effect was noted from the oil spills. One theory is that the biota of the California coast may have a high tolerance to oil as a result of natural seepage over long periods of time. Unusually large amounts of sediment and debris washed into the channel by heavy rains could have acted as a sinking agent for the oil, or the presence of oil may have resulted in a high population of oil-degrading bacteria. In any case, Straughan reported almost no damage to the fish and plants in the area. Dr. Straughan's report was severely attacked as an oil industry whitewash by several advocates of the insidiousness of oil on water Notable among the attackers was Dr. Max Blumer, a Woods Hole Oceanographic Institution chemist, active in the Environmental Defense Fund, who questioned Dr. Straughan's thoroughness and partiality.

Straughan's findings were corroborated in a later report by the Water Quality Office (WQO) of the EPA. The WQO report, entitled "Santa Barbara Oil Spill: Short Term Analysis of Macroplankton and Fish" (1971), was produced by an agency that historically has played up the bad effects of oil spills since their appropriations to some extent are based on public and congressional outrage regarding spills. Among the conclusions of the WQO report are the following:

> About one month after the blowout, midwater trawl collections of macroplankton (small fishes and invertebrates) from different depths in the Santa Barbara Channel and the Santa Cruz Basin showed no significant decreases in species diversity, evenness of abundance, overall abundance, or increase in patchiness, relative to similar collections from previous years. Deep and shallow macroplankton communities either showed little change or showed increased diversity and greater abundances of some species in the Channel. These few detectable changes seemed to be caused by oceanographic anomalies, rather than to the

oil whose effects should be detrimental. Strong off-shore winds with severe winter storms ventilated the basins. Also, deep intrusions of water from the south may have affected the distribution and abundance of a few deep-sea species. Shallow collections contained apparently healthy fish and crab larvae.

The shallow bottom fish fauna about the "oiled" kelp beds off Santa Barbara was more diverse and abundant than its counterpart near Zuma Beach, where the environment appeared to be less productive. Collections made off Santa Barbara in 1969 generally did not differ significantly in abundance from those made in 1967 before the spill.

Collections of nearshore bottom animals trawled off Santa Barbara in spring, 1967, resembled those trawled after the blowout in 1969. Statistical comparisons of fish collections between the two years revealed only 5 significant differences in 11 possible contrasts. These differences were attributable to but two extraordinary collections, one from 1967 containing large numbers of rockfishes, the other from 1969 containing an unusual diversiy of rare flat-fishes. These were unlikely results of oil damage."

In seeming contradiction to the Santa Barbara findings, Blumer (1971) reported a severe kill of littoral life due to a spill in a cove near West Falmouth, Massachusetts, from a grounded tanker. The small tanker carrying fuel oil was caught in an Atlantic gale that blew for two days from the southwest The storm churned up the spilled oil in storm waves and buried oily sediments up to a depth of 10 meters in some places along the shore. Since oil does not readily degrade under anerobic conditions, the oil remained buried in the sediments until slowly shifting sands allowed it to seep to the surface and spread to nearby marshes. Therefore the effects of Blumer's Falmouth spill were unusually prolonged.

Dr. Blumer tends to draw grave conclusions rather easily about the affects of oil spills. In 1969, he wrote:

' While the direct causation of cancer by crude oil and crude oil residues has not yet been demonstrated conclusively, it should be pointed out that oil and its residues contain hydrocarbons similar to those in tobacco tar.

Here Blumer appears to be implying that crude oil can cause cancer: a disease whose mere mention frightens millions of laymen. The carcinogenic hydrocarbons in tobacco must be applied at relatively high temperature to sensitive tissues of the human body for years in order to cause cancer, and the vast majority of smokers still do not contract cancer despite extended periods of smoking. The probability of someone's developing cancer from crude oil spilled on the ocean is so remote as to be ridiculous. Oil spilled on the ocean rapidly spreads and disperses. It is at the ambient temperature of the surface waters. A susceptible individual would have to drink tons of heated seawater over a period of years in order to achieve the same dosage of hydrocarbons as inhaled by the average smoker in a month. There is no question that some fractions of crude oil in sufficient quantities are toxic to man, but Blumer's statement on cancer goes much beyond that.

The concern with the deleterious effects of petroleum oil spills on water is justified to an extent because some oil fractions are obviously toxic in some cases to some organisms. Both Blumer (who reported that a spill confined to a littoral region kills marine life) and Straughan (who found that an open-sea spill dissipates rapidly and appears to have no harmful effects) are correct. There is really little conflict between the positions of Blumer and Straughan.

Since oil spills near the shore are often harmful, the sensible thing to do with oil is keep it away from the shorelines and wetlands. This can be accomplished to a great extent by use of offshore storage and offshore moorings; that is, terminals located far out to sea which allow the oil to be piped to shore facilities by underwater lines.

When spilled on the sea, oil loses its more toxic, volatile components and does not appear to have a lasting, measurable, deleterious effect on plankton or fish, though there is some concern that sea-going mammals may be affected. Oil may taint fish and bivales, giving them a disagreeable taste.

Crude oil spills are relatively shortlived and even without help from man virtually disappear in a few weeks. Most of the oil that has spread will evaporate, be decomposed by bacteria primarily into water and CO_2, dissolve in the sea water, and quickly wane away. The actual remnant time of the oil depends on the type of crude oil, the temperature of the water and the nature of the sea. For example, Lybian oils with a low content of residuals (that is, substances boiling below $400°C$) will evaporate more rapidly than the sulfate oil of Kuwait. Certain oils, such as Venezuelan oils with high values of asphaltenes, spread and are emulsified more readily than others.

Microbiological degradation of crude oil increases with temperature, but even near freezing crude oil conversion to carbon dioxide is rather rapid. Microorganisms that feed on oil are most common in coastal regions where oil spills are more common.

The technical literature indicates quite clearly that certain types of oil spills near shore areas will kill many types of littoral marine organisms, but that oil spilled on the open ocean where it is rapidly spread and dispersed by the wind, waves and currents is not harmful to fish and other marine organisms with the exception of sea birds who happen to land in it.

Perhaps the ultimate proof that spills in open water do little, if any harm to ocean life has been supplied by nature itself. In a number of offshore areas, natural seepage of submarine oil occurs in fairly large quantities. The seepage amounts to about 100 barrels a day (compared to the 30 to 100 barrels per day leaked by the Union Oil mishap) and has been noted by VanCouver and other explorers since the 17th century. This continuous flow of crude oil into offshore waters has resulted in no discernible decrease in marine life in the area.

In addition, many estuaries with heavy oil pollution loads are rich in marine life. New York Bay and the Hudson estuary supports a large and varied fish population, which is encouraged by a rich infusion of waste human nutrients, despite more than 500 oil spills a year.

AESTHETICS OF OIL SPILLS

Oil on water is often ugly, but this has little to do with the myth of ocean destruction due to oil spills. The major complainers after the Torrey Canyon spill were the business people who were afraid of losing beach-bathing customers. Much of the noise in Santa Barbara was made by the local business people worried about the cleanliness of the beaches, not about life in the surrounding sea. Conservationists partial to sea birds also are often quickly galvanized into action by spills. But, as Straughan's and the EPA studies showed, little damage was done to the local eco-systems in the closely monitored Santa Barbara spill, and what small damage was done there was temporary.

An interesting example of the reaction to an oil spill in a populated harbor occurred in July 1971, when a sailor opened up the wrong valve of the U.S. Navy ship Towles in New York Harbor. Thirty-eight thousand gallons of fuel oil from the Towles' tank drifted over to Brooklyn's Coney Island beach, which caters mainly to the city's low-income residents. Local beach businessmen went on T.V. to announce that they were suing the Navy for one million dollars a day for every day there was oil on the beach. The mayor and many other city officials appeared on television to comment on the misfortune. The head environmental politician in New York estimated the cost of cleaning up the oil at $350,000 per day. Meanwhile, youngsters fished for crabs and porgies from the Steeplechase Pier in the middle of the blackened beaches. Despite the hullabaloo, there was no apparent damage to local marine life.

WHAT IS BEING DONE TO PREVENT AND CLEAN UP OIL SPILLS?

The oil industry estimates that it spent more than $100 million during 1970 to combat and control oil spills in

132

an effort to clean up a bad public image, and because the losses of oil cost money. A survey conducted by the American Petroleum Institute indicates that the industry's expenditures for environmental conservation have more than doubled in the past five years. Details are published in a pamphlet, "Report on Air and Water Conservation Expenditures of the Petroleum Industry in the United States, 1966-1970."

When it is desirable to disperse the oil rapidly, emulsifiers and solvents can be used. These chemicals are usually sprayed onto the surface. Agitating the water surface, either with a screw or a powerful jet of water, disperses the oil into extremely small droplets. Each drop, covered with emulsifier, cannot coalesce with the others and eventually disperses into the sea. This treatment is valuable for minimizing damage to beaches. But while most oil has little effect on ocean life, the emulsifiers commonly used to disperse oil are often toxic to marine life. For example, shellfish succumb to detergent concentrations between 10 and 300 parts per million (ppm) of sea water. Smaller marine invertebrates die at concentrations as low as 1 ppm.

Exotic chemicals with names like Polycomplex, derived from a hydrocarbon, Corexit, and Dispersol OS, have been compounded to break down the composition of oil until it can dissipate in the sea.

Mechanical processes developed to handle spills range from carbonized sand sprays intended to sink the oil, to floating fences called "booms" which haven't worked well in heavy seas. The U.S. Government has developed a system using rubber or plastic bags that can be airlifted to a stricken ship, filled with oil from the tanks, and then towed to shore for salvage. Researchers are trying to find better absorbent materials than the traditional bales of straw to soak up spilled oil. One, a polyurethane foam in granular form, looks promising. Specially developed ships that pick up the oil and then separate the water from the oil have been used with some success in calm water. But the present technology for handling major spills is still, on the whole, ineffective and expensive.

Legal measures to prevent oil spills are growing. Nations are adopting stringent laws to dissuade careless or deliberate polluters.

In September 1971, the U.S. Senate ratified a treaty dealing with oil spills on the high seas, beyond a country's territorial jurisdiction. The treaty allows an endangered coastal country to take action against a vessel leaking oil. The action includes blowing up the vessel. The International Convention Relating to Intervention on the High Seas in Case of Oil Pollution, and amendments to another international oil compact, passed 75 to 0.

U.S. Public Law 91-22 requires that vessel owners have financial responsibility coverage equal to $100 per gross ton of their vessel up to a ceiling of $14 million dollars to reimburse the cost of cleanup of oil spills. A vessel not in possession of a Federal Maritime Commission certificate of financial responsibility is not permitted to navigate U.S. waters or enter U.S. ports. This federal law implements a federal contingency cleanup plan in cooperation with state governments and local port and harbor authorities to assure quick cleanup and abatement of oil discharges.

Other U.S. federal laws protect against discharges of contaminants in federal and state navigable waters:

1. The Refuse Act of 1899 (33 USC Section 407) prohibits deposit of any refuse matter in navigable waters of the United States and includes penalties for a violation.

2. The Act (33 USC Section 441) relating to harbor waters in New York, Virginia, and Maryland prohibits discharge of refuse into these waters and prescribes various penalties.

3. Acts implementing the pollution convention of 1954 and the 1969 amendments thereto, (33 USC Sections 1001-15) forbid operational discharges from oil tankers.

In 1972 the U.S. Senate will consider the Waterways Safety Act along with proposed amendments to the 1936 Tanker Act to strengthen construction requirements for U.S. tankers, already the most stringent in the world.

The Water Quality Improvement Act of 1970 forbids willful oil spillage and provides that anyone spilling oil must report it immediately and clean it up. The U.S. Coast Guard may do the cleaning itself and then bill the party responsible. Penalties of up to $10,000 may be imposed on willful spillers. About 500 spills were reported during 1971 in New York City waters, of which 60 percent were of unknown origin, so-called "mystery" spills. Most were minor, averaging less than 25 barrels. Commander Robert Hanson of the Coast Guard's New York office estimates that it costs $10 to $15 per gallon to clean up an oil spill in and around New York.

EXPECTATIONS FOR THE FUTURE

Oil prevention and cleanup techniques have improved greatly within the past three years and will without doubt decrease the damage that is bound to result from accidental spills. However, it is noteworthy that accidental spills play a relatively minor part of the overall oil pollution of the oceans.

Some 550,000 tons per year are spilled into the ocean when tanker operators clean their tanks. Of this, 500,000 tons comes from "dirty" operators, mainly Greek, who operate about 20 percent of the world's fleet. The oil shipping community is presently pressuring the "dirty" owners to use the relatively inexpensive "load-on-top" technique. Washings from all the tanks are stored in one tank and are not discharged overboard.

Newer tankers are being built with double bottoms to prevent oil spills even when the outer skin is pierced

The world's merchant marines would prefer an international agreement covering liability and cleanup of oil and other polluting discharges, since it would eliminate the messy requirement of trying to comply with a plethora of conflicting national, state, and local laws in coastal nations around the world. Tankers are being fitted with oil separation equipment at great expense to cleanse bilge and tanker wash waters of oil. I have been working on a membrane system that will keep the ballast water always separated from the oil in tankers and in the growing number of offshore oil storage vessels. Some of the oil companies are also investigating flexible membranes for oil tankers.

Most of the major oil transportation companies have banded together to provide insurance against damages caused by oil pollution. In 1971, some of the companies taking part in the TOVALOP (Tanker Owners Voluntary Agreement Concerning Liability for Oil Pollution), a group that compensates up to $10 million for oil spill cleanup, have formed a voluntary fund to provide supplementary compensation, bringing the figure up to $30 million per incident. The companies participating in the fund collectively handle over 80 percent of all oil moved by sea. Participation in TOVALOP is open to all oil companies.

Many oil slicks would be eliminated if shippers used larger and fewer oil tankers and the tankers were unloaded at offshore terminals away from the littoral. The probability of oil spills would be considerably reduced in congested harbor regions since one 200,000-ton tanker offshore is less likely to be involved in a collision than four 50,000-ton tankers.

The conditions above are interrelated because it is impossible for a fully loaded 200,000-ton-or-larger tanker to come into any existing American port. American oil companies presently fiddle around with relatively small tankers that require more frequent visits to our harbors.

Every major industrial maritime nation except the United States uses offshore single-point moorings, that hold tankers in deep water while allowing them to pump their cargoes to shore by way of underwater pipelines. In the United States the use of single-point moorings is being held up by rivalries between ports, by politicians who fear that such a mooring might allow oil to spill near their summer homes, by ill-informed environmentalists, and by American oil companies who have tied up capital in antiquated oil terminals, barges, and tankers.

Oil spilled on the 95 percent of the ocean that is a relative biological desert can be beneficial since bacteria break down the oil droplets and zooplankton eat the products, which are nutrients. It is only in enclosed nearshore areas and estuaries that oil spills become deleterious. If by the use of offshore moorings the oil can be kept away from shore area, nature would take care of open sea accidental oil spills and the harmful effects of spills would be reduced to a minimum.

Oil has given man three boons: warmth, power, and mobility. Pollution, through mishandling and improper use of oil, is one price that man up to now has paid for these boons. As with most pollutants, it is the rich who benefit most from the boons brought by oil, and the urban poor who suffer most from the pollution caused by its use. The demand for power and mobility for all will undoubtedly increase, but fortunately man is now in a better position to cope with pollution problems resulting from increased use of oil.

8

FROTH AND FOAM IN THE DETERGENT WARS

> ...the quality of cleanliness. Can anything be more
> entirely artificial? Children and the lower classes of
> most countries seem to be actually fond of dirt: the
> vast majority of the human race are indifferent to it.
> John Stuart Mill: Nature

One of the hottest environmental questions in the pollution wars is whether phosphates in detergents are harming the quality of water in rivers, lakes, and oceans. Confusion abounds as experts testify for and against the continued use of phosphates. The detergent industry is strongly committed to the use of phosphates in household detergents, arguing that substitutes are either dangerous, more expensive, or both. A number of environmentalists argue that industry's main concern is not health, nor the environment, but higher profits.

Since 1948 detergents have flushed soaps from American sinks. Since 1948 soap use has decreased from about 2.5 billion pounds to about 1.0 billion pounds in 1970, while synthetic detergent use has foamed up from less than one-half billion pounds in 1948 to over 5 billion pounds in 1970. Overall, the volume of cleansing agents has about doubled in the past 22 years, although the U.S. population has increased by less than 40 percent, a curious fact to begin with.

There is little question that phosphates, being an essential nutrient, contribute to the growth of algae in some bodies of fresh water. In some lakes and rivers excess algae growth may deplete the oxygen supply and thus prevent the growth of other marine organisms. The detergent question revolves about the advisability of banning a useful product that has moderately undesirable side effects and replacing it with others that may be more expensive and harmful in other ways.

Detergents first appeared on the market in the 1930's, but were little used until the 1950's, when they replaced soap flakes and powders. Detergents were especially popular with housewives in hard-water areas. Old fashioned soap flakes are not well adapted to dishwashers and clothes-washers; they leave a scum in the washing machine and are impractical for use on synthetic materials, designed to be cleaned only by the newer chemical detergents.

As the first detergents superceded soap in the early 1950's, people began to notice foam and billowing suds in the waterways. Suds pushed sewage workers out of treatment plants; people using well water often drew water from their taps with a head on it; white foamy mounds washed up on the shores of lakes and streams.

It took the soap industry six years to find a substitute for the ever-foaming detergents, and nine years before the new product was introduced, after some nudging from state and federal authorities. In 1965 the industry switched to biodegradable detergents with sudsing agents that bacteria could break down. About that time the phosphate-based detergents were first accused of causing water pollution. In 1967, because of fears that phosphates flushed away from detergent washings were harming water-ways, these detergents were discouraged by the government and substitutes sought. Since 1967, phosphate detergent use has been severely limited in many states and banned outright in some localities.

At the height of phosphate-detergent use, domestic sewage contained 3 to 5 pounds of phosphorous per person per year, of which about half came from detergents. Human waste contains 1.0 to 1.2 pounds of phosphates per person annually, a quantity that varies with diet. Typically, detergents account for 30 to 50 percent of the phosphates in municipal sewage, phosphates being the form of phosphorous most likely to cause unwanted algae growths. But not all the phosphorous can be used by plants.

During the period 1957 to 1968 phosphorus in municipal water supplies increased by 40 percent simply because more people were discharging into more sewer systems. This increase had little to do with phosphate detergents.

On September 15, 1971, U.S. Surgeon General Dr. Jesse L. Steinfeld said: "My advice to the housewife at this time would be use the phosphate detergent. It is safe for human health." On October 1, Dr. Steinfeld said to the Subcommittee on the Environment of the Senate Commerce Committee:

It is an oversimplification to equate the removal of phosphates from such products (detergents) with an end to eutrophication and the death of our lakes and streams. Manufacturing, chemical fertilizers and human wastes are other major sources of this chemical. The problem is not phosphate detergents; the problem is eutrophication of our environment.

Appearing before the same subcommittee, Environmental Protection Agency Administrator, William D. Ruckelshaus said:

The rate of eutrophication is controlled by the availability of nutrients such as phosphorus and nitrogen. Phosphorus does not appear to be controlling in most of our coastal waters. We believe phosphorus is the controlling element for this process in most of our country's fresh water resources. Again, we do not know exactly how many bodies of water are controlled by phosphorus, but some of the most pronounced examples of eutrophication, such as Lake Erie, are situations in which phosphorus is the controlling nutrient.

But few environmentalists were listening. The attack on phosphate detergents continued. There were pained cries from many good people who thought they were fighting the good fight to save the environment. The New York Times commented editorially:

The federal government succeeded last week in undoing several years of public education on the harm that detergents do to the nation's waters. In an action as unnecessary as it was sudden and confusing, four

140

of its top health and environmental officials urged a return to phosphate detergents on the ground that alternative cleansers were worse.

Nothing in the government's explanation justified its panicky reversal. Its explanation, it may be pointed out, followed closely the line of reasoning advanced by leading elements in the soap and detergent lobby. Whether or not it was industry pressure that accounted for the change, people who have been accurately warned of how much damage phosphates can do to lakes and streams will find it difficult to take on faith the next warning that issues from the guardians of the environment.

The editorial writers in this case (as well as other environmental matters) apparently do not read the news columns closely.

NUTRIENTS, PLANT GROWTH, AND CULTURAL EUTROPHICATION

Plants require vitamins and nutrients to grow. The major nutrients for most green plants, whether aquatic (algae) or land-based, are carbon, oxygen, nitrogen, phosphorus, sulfur and at least a dozen other trace elements, among them magnesium, selinium, and zinc. The lack of any one necessary element will prevent the plant from growing. For proper growth in aquatic plants there exists a definite ratio of essential elements. Nitrates to phosphates must be present in a 10 or 15-to-1 ratio. It does a plant no good to double the amount of phosphates since the plant can only utilize the substance in the fixed ratio of 10 or 15-to-1.

A nutrient does not limit growth if, when it is increased, no effect on growth is observed. Such lack of growth limitation may be caused by either the nutrient's being already present in optimum amounts, or by some other limiting factor. This is the Law of the Minimum, first postulated by Liebig. Simply stated, the law decrees that growth of an organism is limited by whatever essential factor is in shortest supply. The ratio of nutrients is as important as the quantity of nutrients.

Phosphates, along with nitrate runoff from land are a cause of increased algae growths in many fresh water bodies. Natural eutrophication involves an increase in the biological activity (productivity) of a water body as a result of natural nutrient enrichment. Man-induced growth in lakes and rivers is known as "cultural eutrophication."

In most coastal marine environments, nitrogen, not phosphorus, is the limiting factor in overgrowth of algae. John H. Ryther and William M. Dunstan of the Woods Hole Oceanographic Institution, in Woods Hole, Massachusetts, found that in the surface waters of the ocean there is always an excess of phosphate, relative to the amount of nitrogen available for nutrient use by algae. Ryther and Dunstan determined that although phosphates may be the critical nutrient in many freshwater lakes, in coastal waters and salt estuaries, "it is unquestionably nitrogen that limits and controls algal growth and eutrophication." Removing phosphates from detergents would be pointless in the coastal waters that receive roughly half of the country's sewage. Replacing phosphates with a nitrogen-containing compound (such as NTA) would probably increase eutrophication of the nation's coastal waters. As Ryther said, it would be "only adding fuel to the fire." The situation in many freshwater lakes and streams is different: the consensus holds that phosphates cause unwanted algal blooms.

However, some scientists and detergent manufacturers even dispute the generally accepted notion regarding phosphate-caused growths in fresh water. Dee Mitchel of the Monsanto Company said:

> On the basis of these results, it is evident that domestic waste water will produce eutrophic conditions in receiving waters. However, the data are not in support of the often stated position that the simple elimination of the phosphates from detergents will significantly decrease the rate of eutrophication caused by the resulting waste waters. Furthermore, the data show that the "ecologically safe," high-alkalinity, high-carbonate detergents offer no improvements.

Amid the froth and foam the cry went up for phosphate-less cleaning agents. Valiantly attacking sellers of phosphate detergents, on January 18, 1972, Henry L. Diamond made page one of the New York Times. Mr. Diamond, head of the New York State Department of Environmental Conservation, leading a group of reporters and photographers, swept down on unsuspecting New York City grocery store operators and fined those grocers whose shelves still carried detergents with more than 8.7 percent-by-weight of phosphorus. David Bird of The Times reported that Diamond told one grocer: "Remove the high phosphate detergent...its going to help the water." Presumably he meant the saline water around Manhattan.

I wrote a letter to Diamond asking him if he was aware of studies which showed that phosphorus is not a limiting factor of algae growth in saline waters. Donald Stevens, Director of the Bureau of Water Quality Management, replied to my letter by skirting the question entirely. I sent off a more direct letter to Stevens which read in part:

> My question relating to phosphate-detergents, directed to Commissioner Diamond, was prompted by reports and pictures of him roving about New York City and fining shopkeepers who happened to have detergents on their shelves with more than 8.7 percent elemental phosphorus In light of the findings of Ryther and Dunstan (Science, March 12, 1971) it would seem that Commissioner Diamond is wasting his time, because the amount of phosphates flushed into New York's saline estuarine waters has little effect on the quality of the water. If you have information that contradicts Ryther and Dunstan I would appreciate seeing it. If you have information that would indicate that the phosphate content of New York Harbor or Long Island Sound can be changed by limiting detergents to 8.7 percent, so that these waters will be fit for some superior public use, I would very much like to see it.

Stevens was more forthright in his reply to this letter. He wrote:

> Most of our saline waters are normally rich in terms of the phosphorus needs. They also seem to inhibit the growth of nitrogen fixing bacteria that are common in fresh water. As a result, nitrogen is more of a limiting factor in saline waters as opposed to phosphorus in fresh waters.
>
> Once again I must point out that the law which we must enforce does not make a distinction of this type and there is no room for judgement.

His last phrase aptly characterizes much that is going on in the name of pollution control.

SUBSTITUTES?

When told to find a substitute for phosphates, that is, compounds of sodium, oxygen, and phosphorus that have been used as a major ingredient in detergents since 1948, the detergent industry started looking around for other chemical cleansers. The chemical most commonly used, sodium tripolyphosphate, softens water (that is, neutralizes mineral impurities) and prevents washed out dirt from being redeposited on clothing.

One of the first replacements put forth by the detergent manufacturers was NTA, sodium nitrilotriacetate, introduced in the late 1960's as a chemical substitute for phosphates in detergents. NTA is free of phosphorus, and is regarded as being even better than phosphate in its ability to soften water and prevent redisposition of dirt. But NTA contains nitrates, which, as Ryther and others have shown, when introduced into coastal waters, can cause greater growth of algae than phosphates. In addition, there is reason to believe that NTA combined (chelated) with mercury or other heavy metals can cause birth defects. U.S. Surgeon General Jesse Steinfeld reported that studies by Department of Health, Education, and Welfare (HEW) scientists of rats fed NTA indicated that it was teratogenic. NTA may even be carcinogenic, and the Attorney General has asked the detergent manufacturers to prove that it is not.

Caustic or washing soda, known chemically as sodium carbonate, is commonly found in soaps. Caustics are not as effective as either phosphates or NTA in cleaning clothing, especially in areas having hard water. Caustics can damage eyes and mucous membranes. Children are prone to swallow powders kept about the house, and indeed some have died painfully from ingestion of caustics.

Enzyme products were added to some detergents as a substitute for phosphates. Enzyme detergents contain an active substance derived from a bacterum, Bacillus sub-tilis. Unfortunately, the enzyme detergents contain biological material that can sicken some sensitive individuals. Dr. René Dubos in an article in Science pointed out some of the potential dangers of enzyme detergents. A few cases of allergic reactions to cleansing products were reported before the cleansers were quietly pulled off the market.

Soap has been heralded by some concerned people as the perfect (environmentally speaking) cleaning agent. In soft water areas soap cleans lightly soiled articles with no trouble But in hard water areas the water must be softened either with caustics or detergents. Soaps leave a scum in washing machines and inactivate flame-proof finishes on clothing. Besides, soap costs more than detergents. What is worse from an ecological point of view, soap being organic in origin, contributes to the demand for oxygen just as does unwanted, decaying algae. If America switched to soap, the Biochemical Oxygen Demand (BOD) in our waters would increase by as much as 25 percent and have a damaging effect in many lakes and rivers.

About 50 percent of the U.S. population lives within 50 miles of salt water. Forced replacement of phosphates appears not only ecologically foolish but also costly to the millions of consumers living near the sea: soap currently costs 20 percent more than detergents and its price may rise as it becomes scarcer, and non-phosphate detergents are not effective cleansers.

HOGWASH, WHITEWASH

It is obviously silly to ban the use of detergents in coastal communities on the grounds that it causes harm to estuarine waterways, when this is simply not the case.

Phosphates can be removed from sewage at a relatively small cost. They can be precipitated as part of a process that also improves removal of other undesirable toxicants and chemicals from the waste treatment plant effluent. Sewage precipitation costs are comparable to the anticipated cost to the consumer of new non-phosphate detergents. Detergent makers estimate that replacing the 2.5 billion pounds of phosphate detergent would cost the consumer $200 million to $400 million annually A federal official estimated a public cost of upgrading sewage plants should be as much as $500 million (September 17, 1971). Whether manufacturers would make more money one way or the other is debatable. According to Barry Commoner: "Detergents that pollute produce nearly twice the profits of soaps that don't... New, more polluting technologies yield higher profits than the older, less polluting technologies they have displaced." As evidence for his conclusion, Commoner says that the soap industry increased its profits from 31 percent of sales in 1947 to 54 percent in 1967 by emphasizing the sale of detergents over soaps. Commoner ignores the fact that volume of cleansing agents sold had doubled during that period, and tht inflation alone could account for much of the price and profit rise.

In this great land a large aggressive national company could make a profit by selling garbage to a large segment of the American television public. Some companies smelling profit in Commoner's cause hit the market with soap advertised to be ecologically beneficial or having "ecology" worked into their name. Such high sounding names as Ecolo-G, Pure Water, Un-Polluter, hit the housewife in her environmental concern and in her pocket book, capturing at their peak some 10 percent of the market. These products were invariably more expensive than phosphate-based soaps. Some of them were subsequently banned because they contained hazardous substances. The gullible housewife is continually bombarded with products that are "new," "whiter than white," "amazing," "washday miracles," "improved," "fabulous." The last adjective most accurately serves the whole industry and is based on the fable that cleanliness and super white sheets are close to godliness.

Proctor and Gamble, the largest American merchandiser of soaps, reported record earnings of $276 million for the year 1971. In response to local laws against the sale of phosphate detergents, the company had stopped selling laundry detergents in the Miami, Buffalo and Chicago areas. These actions did not seem to effect overall profits.

Ronald O. Ostrander, the project engineer who developed the best-selling detergent Tide for Proctor and Gamble said that a "great hoax" has been foisted upon consumers by encouraging them to use more of a detergent than needed for a clean wash. He told the House Conservation Subcommittee that one-tenth of the recommended level of Tide will give the housewife a "clean, bright, sanitary laundry with minimum of color fading and loss of fabric tensile strength."

While several U.S. states and counties have adopted laws restricting the sale of phosphate detergents, Suffolk County on Long Island is the only area in the country thus far that has banned the sale of all detergents. Suffolk draws all its drinking water from wells, and since 95 percent of the county uses cesspools (septic tanks) the ban on detergents was instituted to preserve the water supply for the 1.2 million inhabitants of this growing suburban area. The New York Times (January 17, 1972) interviewed 17 women at random; four said "they objected strenuously to the ban and wanted to see it lifted; four said they didn't like using soap but would abide by the law; and nine women said that they didn't really notice the difference." Some of the women said they "believed soap was cleaning their clothes better than detergents."

The individual consumer can save money and perhaps persuade himself of having benefited the environment by bathing and washing less frequently. That would justify my feelings as a child, that excessive bathing somehow removed naturally protective substances from my body.

147

CLOSING THE CIRCLE BY
DUMPING WASTES INTO THE SEA

"Human beings have broken out of the circle of life, driven not by biological need, but by social organization which they have devised to 'conquer' nature: means of gaining wealth that are governed by requirements conflicting with those that govern nature. The end result is the environmental crisis, a crisis of survival. Once more, to survive, we must close the circle. We must restore to nature the wealth that we borrow from it."

Barry Commoner: The Closing Circle

Man has been using the ocean as a receptacle for his wastes since Adam first ate that apple; but the sea, covering 70 percent of the earth's surface, has nevertheless changed little. The ocean is a prime source of man's protein food; it determines the balance of oxygen and carbon dioxide in the atmosphere; it controls global climate and it is the prime source of water vapor that falls as rain on the continents. The ocean is used by those people fortunate enough to live near the littoral as the ultimate global container for solid and liquid wastes throughout the world. In the United States solid wastes, including sewage sludge, expected to triple in volume by 1999, and use of the ocean for disposal is expected to grow. Yet many people are adamantly against ocean disposal of wastes, and an environmental dispute has arisen regarding the ecological wisdom of dumping contaminants at sea.

Rational discussion of the condition of the ocean and the effect of ocean dumping has been muddied by lugubrious pronouncements. For example, Dr. Jacques Piccard has said that "at the current rate of pollution there would be no life in the oceans in 25 years," and politicians have claimed that dumping of sludge and spoil in the New York Bight has created a 20-square-mile "dead sea."

In November 1971, the Senate passed a new water pollution control bill by a vote of 86 to 0. The bill would in effect outlaw discharge of all pollutants into U.S. waterways by 1985. The bill prohibits certain dumpings within the three mile limit and requires permits for all other ocean disposal. A strong compromise water pollution control law was passed by both houses of Congress in the fall of 1972.

The Senators' great interest in the ocean stems, in part, from the widespread publicity given to gloomy pronouncements on the sad condition of the sea from Jacques Cousteau, Thor Heyerdahl, Barry Commoner and others. In front of intrigued Senators, Cousteau repeated his mind-boggling pronouncement that man has damaged 30 to 50 percent of life in the sea. Heyerdahl reiterated his headline catcher concerning his raft trip across the Atlantic in which he claims to have observed oil clots floating on the surface of the Mediterranean Sea and the Atlantic Ocean on 40 out of his 57 days at sea.

Cousteau's vague pronouncement is incredible. Marine species have changed little if at all in the past two hundred years. Today, though, a few aquatic mammals are in danger of extinction, and overfishing on a few fishing grounds threatens to decimate certain stocks of marine species, the overall productivity of the ocean, if anything, has increased due to increased man-induced runoff of nutrients into the sea. This is made evident in part by the doubling of the world's catch of fish during the past 0yrs. Heyerdahl's famous oil clots are to be expected in the landlocked Mediterranean, which has in recent years seen a tremendous increase in oil shipments from Libya, Egypt and Algeria to Europe and the Americas. Oil clots would also tend to concentrate in the Sargasso Sea, through which Heyerdahl passed, as it is the center of the North Atlantic gyre, the clockwise pattern of currents that includes the Gulf Stream and thus is a sink for floating debris from most of the North Atlantic Ocean. In any case, the mid-Atlantic is much less productive than coastal areas, and

149

any nutrients supplied by oil and other biodegradable substances would probably be welcome to the plankton and fish in the area. If Heyerdahl had taken the trouble to note and describe the percentage of time at sea in which he was in contact with oil clots (rather than the number of days) his statement would have been more meaningful and less dramatic.

Underlying the myth of ocean pollution is the mystical assumption that the mighty ocean is somehow sacred and should not be violated by human excrement. Strong support for this point of view has come from the Water Quality Office of the EPA and is expressed in the federal report on ocean dumping:

> Ocean dumping of undigested sewage sludge should be stopped as soon as possible and no new sources allowed. Ocean dumping of digested or other stabilized sludge should be phased out and no new sources allowed... (Council on Environmental Quality, 1970)

We must make a distinction between "sanitary" municipal wastes and industrial wastes. Municipal wastes consist mainly of water in which is dissolved and suspended less than 1 percent human excreta, which is non-toxic and biodegradable. Chemical, radioactive and metallic wastes from industry, on the other hand, are often long-lasting, not readily biodegradable and in higher concentrations, toxic. The myth surrounding ocean pollution from wastes would have it that Americans must not use the ocean for disposal of sanitary wastes or great harm will come to the sea.

Since sea water is not used for drinking purposes and is used only moderately for industrial purposes, the fact that one puts relatively small amounts of biological wastes into the sea is not hazardous, from either an ecological or health viewpoint. The sea can assimilate many different wastes because of its enormous volume and its churning, oxygen-filled waters. Properly used, the ocean offers man a practically unlimited resource for waste assimilation at a reasonable cost. And there are valid ecological reasons why man should use the ocean for dispersal of biodegradable wastes.

The oceans can absorb huge quantities of waste without disturbing the ecological balance. Twelve billion cubic feet of sea water are available for the disposal of the wastes from each individual on earth. Sanitary engineers estimate that mixing normal sewage with sea water in the ratio of 1 part sewage to 200 parts water will allow natural biochemical processes to purify the water. The oceans, lakes and streams, as well as the ocean of air, are all natural purification plants.

Pollutants in water disperse through a combination of diffusion and convection. The dominant factor is usually convection flow. For example, wastes from Florida's Gold Coast, dispersed in the Gulf Stream, move along the lower Florida coast at about 5 knots and then sweep across the Atlantic to British shores. At the opposite end of the convection scale are almost totally stagnant salt water bodies such as the Black Sea. On an average, most of the oceans move at about five centimeters (two inches) per second. Also important to pollution dispersal are diffusion processes, molecular transfer and motion due to the translation of suspended matter.

Variable currents and diffusion parameters can create widely ranging conditions relatively close to one another. Within a radius of a few miles of New York City extremely polluted and relatively contamination-free conditions exist in close proximity. It is not unusual for the phosphate content in these waters to range from good to extremely undesirable conditions within a mile or two of each other.

A Food and Drug Administration study stated that 62 million tons of sewage, chemicals, radioactive wastes, acids, explosives, and solid wastes were dumped into the oceans off U.S. coasts during 1970. The study indicates that 13.8 million tons of dredge spoils were discharged into the Atlantic, while 5.7 million tons of dredge spoils were deposited into other sea waters surrounding the United States. Dredge spoils are composed of sand, bottom sediments and muck sucked up by dredges when cutting channels and clearing boat slips. There are over 280 dumping sites off the Atlantic, Gulf and Pacific coasts.

151

Twenty-nine million tons, or 63 percent of total ocean dumping, occurs in the Atlantic; 28 percent of the total dumping area is in five sites off New York City; and 98 percent of dredge wastes placed in the Atlantic is discharged within the 12-mile limit. Fifty-eight percent of the sites are located within three miles of shore.

In addition to dredge dumping grounds, the Atlantic contains sites 100 miles offshore of New Jersey for discharging industrial wastes, explosives, toxic chemicals, ammunition and radioactive wastes. The two principal areas off the East Coast that receive sewage sludge are located 12 miles from New York City and Cape May, New Jersey, respectively.

New York City has been disposing of sewage sludge at sea since 1937. Most of the city's sludge is digested in a process that, when practiced in secondary sewage treatment plants, destroys over 50 percent of the organic matter by microbial activity and renders the sludge nonputrescent.

Communities near the coast usually pump waste sludge into barges or tankers and dump it at sea in areas designated by the U.S. Corps of Engineers under the Harbors Act of 1888 and the Rivers and Harbors Act of 1905. Some communities pipe their sludge to sea.

Approximately 13,000 cubic yards per day of sewage sludge from the New York area is disposed of in the New York Bight, a quantity that will more than triple in the next 10 years as more sewage treatment plants are built and older plants are upgraded. Fecal coliform bacteria counts are used as rough indicators of pollution from human waste. Samples taken directly in the wake of a discharging sludge boat in the bight indicated a high coliform count in excess of 2.4 million organisms per 100 mileliters for total coliforms (Buelow 1968). However, the concentration rapidly decreases to near normal levels in less than an hour. A blanket of sludge solids covers about 20 square miles of the bottom in the vicinity of the disposal area.

Sludge dumping is relatively inexpensive compared to other forms of disposal. According to Commissioner John Peters, Nassau County barges its sludge to sea at a cost approximately one-third that of other disposal methods.

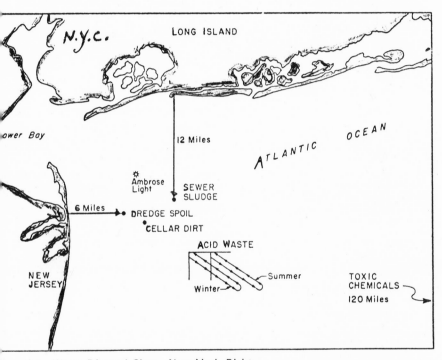

Figure 9-1   Waste Disposal Sites— New York Bight.
Map by the U.S. Army Coastal Engineering and Research Center

New York City, which operates its own fleet of five sludge boats, spends $1.5 million annually for sludge removal and dumping. At a Senate hearing to consider various ocean dumping bills, Commissioner Martin Lang of New York City's Department of Water Resources said that "it is economical; it has cost us on the order of $12 a dry ton, contrasted, for example, with $90 a dry ton in Chicago at this time." Alternate disposal methods for New York, such as incineration and landfill would cost three to five times as much as barging to sea. New York operates 14 sewage plants which treat some 1.1 billion gallons of sewage a day.

Sludge dumping practiced by the city since 1937 has produced no demonstrable unwanted effects in the surrounding shore communities. In fact, there is some indication that the quantity of fish in the New York Bight has increased during the 1950's and 1960's. But since the Hudson River and nearby seaside communities inject vast quantities of waste material, much of it untreated, into the New York Bight, it is virtually impossible to ascertain what nutrients and what pollutants are causing what effect, if any.

SEWAGE TREATMENT PLANTS ON THE EAST COAST

Most of the major cities of the U.S. East Coast are heavily committed to the policy of building expensive sewage disposal plants. Sewage treatment plants separate the effluent flushed from toilets and kitchen sinks into two fluids: treated effluent, doused with chlorine, which is dumped into nearby waters, and sludge, a smelly fluid that is 95 percent water and contains most of the heavier waste particles. In seaside towns the sludge is often piped to the sea or barged a few miles to sea and dumped, where it then mixes with the treated effluent. In essence, present sewage treatment plants separate sewage at great expense into two liquids, which are subsequently mixed into nearby waters

New York City engineers are engaged in an interesting exercise in sewage handling that should cost the taxpayers about $1 billion before they are through. Despite vehement opposition by local residents, city engineers are pushing ahead with plans to build a plant in Harlem on the Hudson River that will cost over $800 million. Operating expenses over a 30 year period will add at least another $300 million. The plant is supposed to treat the fluid waste of the residents on the West Side of Manhattan. The waste now flows untreated into the Hudson estuary. With the plant in operation the Hudson will still be receiving almost as many phosphates, toxic materials, nitrates, and other nutrients it is now receiving spiced with chlorine. Though the river will be slightly cleaner, the decreased oxygen demand of the sewage effluent will have an almost imperceptible effect on the overall quality of receiving tidal waters. August 1971 samplings of Hudson River were taken by the Interstate Santitation Commission indicating good levels of oxygen and low coliform counts in 15 out of 20 stations measured.

On the other side of the Hudson River the communities in New Jersey are building a variety of plants that will, by 1978, treat almost all of the sewage now flowing as primary effluent into the Hudson and New York Harbor. To a large extent the decision to build plants is based on a favorable ecological-political climate rather than sound public health and engineering considerations. In this respect, the colloquoy between Senator Ted Stevens of Alaska and Martin Lang of New York City's EPA is of interest (March 3, 1971).

Senator Stevens: Are you implying that these decisions (to build a one billion dollar plant) that have been made are in fact contrary to the total recommendations you are going to have to work out for the whole area, and you are going to go back and rebuild these?

In other words you are spending a billion dollars that you are going to tear out in two years with federal money?

Mr. Lang: No, we are not going to tear them out, sir. What I am saying is the needs of the future may transcend present day standards and we have to be prepared to move towards them on a rational basis.

Senator Stevens: Why can't you crank the new standards into this plant--this five-year, billion dollar plant? Are we wedded to those decisions made in the past?

Mr. Lang: Well when you start a major engineering enterprise, then it becomes like a stone rolling downhill, you are committed, you have let contracts... In other words there was one point in time when the money was available to build here now. In other words there was a driving urgency, a short timetable to utilize the bond issue. There was no time for introspection, no time for study.

Our nation's sewage systems produce a tremendous amount of waste sludge as a by-product of the sewage treatment process, and society must ultimately dispose of this vast and growing quantity of waste. Sludge and effluents from municipal waste treatment plants can either by lagooned on land, burned to remove the carbonaceous content, dumped into local waters, or sent out to sea. Lagooning wastes is practicable where low land is available, away from residential areas, and where the climate is relatively dry. These conditions rule it out as a useful method for most large cities. Incineration of wastes including sludge is not only relatively expensive, but also often creates air pollution.

After reviewing all of the existing sludge handling and disposal methods in use, R. S. Burd (1968) made the following general observations:

1. Anaerobic digestion followed by sand bed dewatering is the most common method of handling sludge at sewage treatment plants. The obvious reasons for its popularity are simplicity and low cost. Few large cities dewater sludge on drying beds.
2. Lagooning is the most common method that industry uses to dispose of waste sludge.
3. For coastal cities, anaerobic digestion followed by pipeline transportation to the ocean or land reclamation areas is by far the cheapest method of sewage sludge disposal.
4. For many near-coastal cities with navigational access to the ocean, digestion followed by barging is the most economical method of sludge disposal.

5. Marketing dried waste sludge has been generally a failure. Heat drying of sewage sludge is, therefore, rarely given serious consideration by consulting engineers.

6. Sludge treatment presents many operational problems involving odors, inefficient solids capture, constant supervision and general lack of scientific controls.

7. Almost all of the methods of sludge handling and disposal now used were known in 1930.

Burd's observations are borne out by the experience of the cities of Los Angeles, San Diego, and the Hague.

CITIES THAT PIPE WASTES TO THE SEA

Piping of wastes to sea has been successful in California, Holland and elsewhere. The city of Los Angeles disposes of sludge through a seven-mile pipeline to sea from its Hyperion works. The Federal Water Pollution Control Administration (now the Water Quality Office of EPA) would not help subsidize this project, but California engineers felt so strongly about its desirability that they built it anyway. No noticeable pollution has resulted and the results have been encouraging.

The outfall from the Hyperion sewage treatment pipeline has been monitored closely by Los Angeles County. Because no deleterious effects have been found, more long pipelines are being put into place on the West Coast. Orange County recently emplaced a 10-foot diameter, five-mile concrete ocean outfall south of Los Angeles.

San Diego now pipes primary treated waste effluents to the Pacific Ocean. San Diego Bay is a crescent-shaped shallow estuary about 15 miles long and 2.5 miles wide. About 100 tuna fishing vessels, one-quarter of the active Navy fleet, about 700 visiting commercial ships a year, and thousands of recreational vessels use the port regularly.

After unusual blooms of algae in the heavily polluted bay caused by excessive nutrients, San Diego decided to do something about pollution. More than 60 million gallons per day (mgd) of sewage were being discharged into the bay

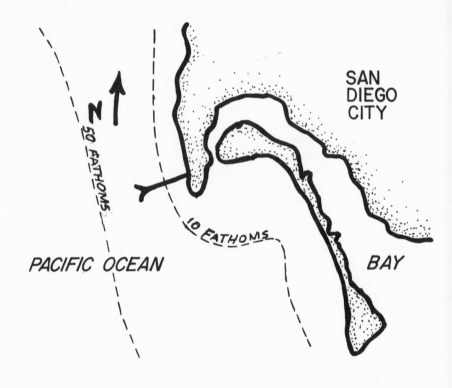

Figure 9-2   The San Diego Ocean Outfall Sewer Pipe.

with little treatment.   A high coliform bacteria count prevented recreational use of the water.

In 1954 a $16.5 million sewer bond issue was defeated,
but in 1960 the people of San Diego approved a $42.5
million bond issue, and construction of gathering sewers
and outfall facilities began in 1961.   The outfall pipe went
into service in 1963.   The total sewage system cost about
$60 million, to which the federal government contributed
only $2.5 million.   The system serves a population of
about one million people, but can handle the wastes from
double that number.   Piping pollutants to sea has resulted
in a sparkling bay, without unsightly and smelly algal
blooms; a bay in which swim sea bass, sharks, and porpoises, as well as thousands of other marine animals.
No deleterious effect on the marine environment has been
observed

An extensive research program monitored the Pacific
environment before and after the effluent pipelines were
placed in service.   Carl Chen, and other scientists, studied
the effects of the San Diego two-mile outfall.   After a
six-year comparison, Chen reported that there has been a
slight increase in ocean life in the area where the pipe
empties at a depth of 200 feet of water. No biological
die-offs have been noted.   Chen and his associates could
detect only minimal changes in water quality occurring in
the water column.   The ambient ecosystems remained
stable, though there was a slight increase in the total
benthic (bottom dwelling) population.

The Dutch, who have a well-deserved reputation for
cleanliness, recently installed a long sewage outfall system
at the Hague.   The area served by the outfall system has a
concentration of industry and more than one million inhabitants.   It was imperative that the sewage system did
not pollute the beaches in the vicinity of a popular resort.

Domestic sewage and industrial wastes are collected
in outlying districts through a sewer network.   The sewage
mixture is then pumped to a central sewage treatment
plant situated below a park, where the heavier sludges are
allowed to settle.   The effluent minus the sludge is piped
out to sea through a nine-foot wide buried outfall pipe and
diffused into the sea water about 8,000 feet from shore.

The heavier sludges are carried six miles to sea through a 14-inch-diameter steel pipe.

When discharged into shallow bays and enclosed water bodies, sludges from sewage plants and other pollutants loaded with bacteria with an affinity for oxygen may deprive some forms of marine life of this essential element. Low oxygen conditions are most likely to occur during the summer, since warm water can hold less oxygen than cooler water. Due to bacterial pollution, over one million acres of shallow-water estuarine U.S. shellfishing grounds have presently been declared health hazards and closed by the Public Health Service. This pollution is due to estuarine and not ocean dumping and would be eliminated if sewage and sludge were piped to sea, rather than into estuaries.

The problem, then, is that man tends to concentrate waste products in rather restricted areas instead of dispersing them where nature can purify them.

The total effect of dumping large and growing quantities of sewage sludge into offshore waters has not been fully determined. Marine organisms, particularly shellfish, may ingest pathogens in the waste sludge that might eventually be consumed by humans. Recent studies have shown that an area of approximately 10 to 20 square miles of sea bottom in the vicinity of Ambrose Light, off the New Jersey shoŗe, has been adversely affected by concentrated, excessive dumping of sludge. Sludge boats from New York City, northern New Jersey, and Nassau County use the prescribed site. There is no longer a natural benthic (bottom dwelling) community in the dump area and bottom sediments contain high levels of bacteria.

Those cities that line the ocean are fortunate. The sea allows them an excellent means to dispose of wastes at relatively low cost, and life in the nearby ocean will benefit from the added nutrients, if the wastes are properly dispersed. For the past few years scientists at the Woods Hole Oceanographic Institution have been using sea water diluted with sewage effluent for culturing natural populations of phytoplankton. Dunstan and Menzel (1971) report that seawater diluted with secondary treated sewage effluent provides excellent enrichment for the maintenance of mixed natural populations of normally occurring coastal

160

phytoplankton. "Before and after" studies of the effect of sewage piped into Chesapeake Bay showed increases in the amount of perch, eel, and sunfish, though some decreases were noted in a few gamefish beloved by fishermen. As the warmed nutrient-rich effluent moved into the region under study, the ecology of the region changed, but the overall effect was an increase in the total quantity of marine protein in the area (Tsai, 1970).

J. P. Wise, a U.S. government fisheries biologist, brought the following facts to my attention (letter of January 24, 1972): Twice as many alewife, a herring-like fish, were caught in the Potomac in 1970 as were taken in 1939. Commercial catches of striped bass in New York waters have doubled during the years 1950 to 1970. Bass use the greatly enriched Hudson River as their major spawning ground in the Northeast. Blue crab catches in most Atlantic estuaries have increased by 50 percent since 1945. Menhaden, a species of commercial fish dependent on estuaries during the early part of their life-cycle, have reached an all time high during recent years in some areas. Landings of menhaden on the Gulf Coast during 1971 were 1.6 billion pounds, the highest landings on record for that area (U.S. Department of Commerce, January, 1972).

These findings hardly indicate that U.S. estuaries and the sea are becoming unfit for marine life, quite the opposite.

CLOSING THE CIRCLE

Sewage and sewage sludge contain nutrients that, when properly distributed, can make the plants in the sea bloom. Dried sludge solids have been used for years as a soil fertilizing agent in Chicago, Schenectady, Houston, Milwaukee, and several other cities which process activated sewage and sell dried digested sludge as a soil conditioner, or use the sludge directly on farm lands.

Sludge contains a spectrum of micro-nutrients that are not found in chemical fertilizers. For example, Milorganite, the bagged, dried sludge derived from sewage sold by Milwaukee, contains these basic nutrients: nitrogen, 6.00 percent; phosphoric acid, 4.59 percent; potash, 0.80 percent;

sulfur (as SO), 1.68 percent; calcium (as CaO), 1.55 percent; iron (as Fe_2O_3) and over a dozen other trace elements, 6.63 percent (Wilson, 1966).

Some examples of existing sewage sludge waste utilization on land are listed below.

In Melbourne, Australia, raw sewage has been used for over 50 years to irrigate and fertilize pasture lands which support over 40,000 cattle and sheep. In Maple Lodge, England, digested sludge is used to fertilize over 2,000 acres of grasslands. The sludge allows the local farmers to keep two-and-a-half cows to the acre against the normal one. In Leipzig, Germany, sewage effluent from a primary treatment plant has been used since 1935 to irrigate 56,000 acres of grassland (Bacon, 1967).

Since 1968, Chicago has been experimenting with use of waste sludge on crops not used for human food. Sludge is taken by tank car to firms in Illinois. The digested sludge has in it 1,000 to 3,500 ppm nitrogen; 500 to 2,000 ppm ammonium nitrogen; 500 to 1,550 ppm phosphorus; 150 to 175 ppm potassium, plus varying amounts of trace metals. Plants fed five to ten inches of sludge grew better than comparable plants fed with convenient chemical fertilizers.

Miami, Florida, provides a good example of intelligent sludge use. In 1956, Miami completed a 47-million-gallons-per-day (mgd) plant that generated about 20 tons a day of sewage sludge. About 30 acres of land around the plant site were planted with Bermuda grass, but there was little topsoil over the sandy fill dredged from the bays, and the grass did poorly. The first application of liquid sludge produced a marked improvement. Continued improvement was noticed with the second and third application of sludge. One man can apply 25,000 gallons of sludge in an eight-hour day. In a matter of weeks, the lawn was ready to be mowed. During 1970, Miami disposed of 150 million cubic meters (4,000 million cubic feet) of digested sludge for about $5 per day per ton (Water Pollution Control Federation, 1971).

It is apparent that waste sludge has been put to beneficial use by communities throughout the world without any resulting health hazard. Sludge disposal in the ocean is likely to result in even less of a health risk than using it directly on food crops. Ocean dumping of selected bio-degradable wastes is in fact highly desirable.

The rivers of the world carry enormous quantities of organics and sediments to the sea, a natural process that has been going on for millions of years, and man's contribution to these wastes is comparatively small. Yet there is no evidence that the chemical composition of the sea has changed for over a million years. The sea has the capacity to assimilate vast quantities of natural wastes that run off from the land, and it seems ecologically unwise not to use it. Man takes millions of tons of fish from the sea. In a way, it is only fair that he balance his harvest from the sea by returning nutrients so that other animals can thrive, and in this way close one vital ecological cycle.

A DASH OF LEAD AND MERCURY

There was a king reigned in the East:
There, when kings will sit to feast,
They get their fill before they think
With poisoned meat and poisoned drink.
He gathered all that springs to birth
From the man-venomed earth;
First a little, thence to more,
He sampled all her killing store;
And easy, smiling, seasoned sound,
Sate the king when healths went round.
They put arsenic in his meat
And stared aghast to watch him eat;
They poured strychnine in his cup
And shook to see him drink it up:
They shook, they stared as white's their shirt:
I tell this tale that I heard told.
Mithridates, he died old.

A. E. Houseman

Recently environmentalists have attacked the use of two heavy metals, lead and mercury, and have brought about the ban of certain foods containing them. A few well documented and well publicized cases of mercury poisoning have led to widespread public fear and some blunt political reactions.

High concentrations of lead or mercury are poisonous, as are high concentrations of almost all known compounds. It is a truism of toxicology that any substance is harmless or deadly depending on how it is administered. Common table salt is poisonous when improperly used; indeed a number of babies in a Binghamton hospital were killed in the 1960's when salt was mistakenly substituted for sugar in their formulas. So it is with lead and mercury; small amounts existing naturally in the environment often find their way into our food and are usually not harmful; large amounts of lead or mercury can kill. The basic questions are: (1) how much is harmful? and (2) are the mercury and lead levels in our environment rising to dangerous heights?

NATURAL METALLIC POISONS

Metallic ions have always been part of the natural environment. Every metal can be found in seawater, and scientists have established that certain metals that are poisonous in large quantities are essential in trace quantities to natural growth. Copper, a poison often used to inhibit marine growth, is essential for reproduction in mammals and growth in plants. Zinc, another poison when improperly used, is driven into orange trees in Florida to promote growth. Quite a few so-called "toxic" heavy metals such as vanadium, mangenese, cobalt, nickel, copper and zinc are, in small quantities, necessary for life.

Under certain natural conditions of soil and climate, some natural metallic elements can poison plants, animals and man. These naturally occurring poisons include selinium, thallium, chromium, lead, and arsenic. Animals often die if they feed in fields with high levels of these elements. Elements sometimes kill by their absence. For instance, the metal cobalt is essential for the growth of cattle and sheep. Over 200 years ago dairymen in Dartmoor, England, noted that their cows pined and sickened when pastured in a certain field, but recovered when moved to another field. This "pining sickness" was later reported in New Zealand, where it was called "Morton Mains Disease" after the district where it occurred. Sheep in Australia died of the disease by the hundreds and could be saved only by driving them to other pastures. American scientists read accounts of this malady in British publications and noted its similarity to events in certain parts of Michigan and Florida. Scientists around the world began to hunt for the cause of Morton Mains disease. After considerable searching, two teams of New Zealand chemists identified a cobalt deficiency as the cause of the ailment. Today cows' diets are supplemented by cobalt, an element that in large quantities would poison them.

165

MERCURY IN THE ENVIRONMENT

Mercury poisoning (argyrism) is not uncommon. However, almost always it results from accidents or deliberate misuse, not from eating common foods. Contrary to common belief, metallic mercury is not poisonous. Children who have swallowed the contents of a thermometer are usually unharmed by the experience. It is mercury compounds, particularly the salts and methyl forms that are harmful, even deadly. While organic mercury compounds are not readily absorbed by the body and are rapidly excreted, methyl mercury, an organic compound, is carried by red blood cells to every part of the body and excreted slowly.

In subacute or low level cases, mercury poisoning leads to chromosome damage, nervous disorders, irritability, insomnia, and constant trembling. This general loss of coordination and peculiar involuntary gestures were the mark of the "mad hatters," who in the 19th century worked in hat manufacturies where mercury salts were used to process felt.

Acute argyrism causes digestive disturbances, excess salivation and mouth ulcers. This is often followed by death due to kidney misfunction unless the patient's blood can be cleaned or an artificial kidney used.

Most mercury poisoning occurs in industry and through accidental use. Two recent events in Japan and one in the U.S. focused the attention of the public on mercury in the environment: the Minimata and Niigata accidents, and the Huckelby family accident.

Minimata Bay, an inland Japanese estuary, for many years had been receiving the mercuric wastes of nearby industrial plants. Fish and shellfish in the area concentrated the mercury in the poisonous form of methy-mercury and the local residents ate a tremendous quantity of the contaminated fish over a relatively long period of time. In 1953, 111 persons fell ill, 41 of them subsequently dying from eating heavily contaminated fish. The mercury was traced to a nearby plastics plant. Rather than close down the plant that was the prime employer in the area, the residents "cured" Minimata disease by simply not eating

fish from local waters. Fish in the area contained con-
centrates of mercury 10 to 40 times the guideline levels
set in the United States by the Food and Drug Administration.
The disease was called Minimata sickness; in 1959 it was
recognized as methyl-mercury poisoning.

In 1964, 26 persons became ill and five died from eating
fish in Niigata Prefecture. The fish were contaminated by
industrial mercurial wastes.

In September 1969, a poor black farmer, Ernest Huckelby,
picked up floor sweepings from a New Mexico granary.
Unfortunately he collected mainly millet seeds, which had
been treated with a methyl-mercury fungicide to prevent
destruction of the seed by fungi when planted. Treatment
of seed in this manner is common in agriculture. The
seed was dyed red as prescribed by law to warn against
eating it. Huckelby, however, fed the seed to 17 of his hogs.
When some of the hogs became ill, he butchered one of
them and fed it to his family. In a few weeks three Huckelby
children fell ill from the meat, which contained as much as
27 parts per million (ppm) of mercury. The three children
suffered severe and partly irreversible neurological damage.
The Huckelby case was widely played up in the press,
complete with pictures of the crippled children.

In the spring of 1970, Bruce McDuffie and other technologists began to measure the concentrations of mercury in
fish in the United States and Canada and found levels of up
to 10 ppm in fish from the industrial Lake St. Claire area
near Detroit. They characterized this finding as "alarmingly
high." The higher levels were similar to those which had
caused Miniwata disease in Japan, and environmental
mercury poisoning became an American environmental
issue.

During the past few years a number of investigators have
gone into the business of measuring mercury levels in
animals and food. The FDA set the permitted level of
mercury in fish at 0.5 ppm (one-half miligram per kilogram),
which eliminated some commercial fishing in a few Great
Lakes areas. The FDA began sampling foods for mercury
and subsequently banned   tuna fish and some Great
Lakes salmon and then swordfish from American grocery
stores. Some fishermen and sport fishing industries have

been forced out of business due to a loss of over $25 million in sales. Also banned were some "health food tablets made from the distilled livers of Pacific seals" which were found to have mercury levels of 36 ppm. Some of the bans have since been rescinded, but the mercury picture is still obscure.

MERCURY IN NATURE AND IN USE

Mercury permeates the natural environment in concentrations averaging 0.05 ppm. Mercury-bearing rocks may contain as much as 30 ppm. As rocks weather, mercury leaches into rivers, lakes and the ocean; a natural process that has been going on since the oceans were formed.

Man adds mercury to the environment when he mines and processes mercuric ores and when he uses mercury in industrial processes. Some of the mercury is washed into nearby waters, as was the case in Minimata Bay and in Lake St. Claire. Processes which require large quantities of mercury are the manufacture of chlorine and caustic soda (23 to 33 percent), agriculture (26 percent) and instruments and switches (23 percent). Mercury is used in such common items as thermometers, mirrors, electric switches, and lamps. It is used by doctors as a diuretic. As much as two pounds of liquid mercury has been fed to patients with no apparent ill effects. Mercury is also commonly used as an amalgram for filling teeth.

But by far the greatest input of excess mercury into the environment comes from the burning of fossil fuels. Mercury in coal ranges from 0.5 to 3.3 ppm. When the coal is burned, the mercury rises into the air with the smoke and later descends to the ground. In the United States, burning of coal spews some 2,000 tons of mercury annually into the environment. This may seem like a lot, but on a nation-wide scale, 2,000 tons amount to less than two-thirds of a pound per square mile per year. But the natural degassing rate of the planet earth, more than 70,000 tons per year, far outweighs man's contribution. There has been no significant rise in mercury levels in the ocean, which by natural processes contains an estimated 100 million tons.

Environmental mercury is innocuous when it is buried in underwater sediments. Specialized bacteria convert harmless mercuric sulfide to toxic methyl-mercury. If layers of sediment cover the mercury-bearing sediments, the poison cannot enter the food chain. Potentially poisonous but quiescent mercuric deposits can be found in the vicinity of most large plants that discharge spent mercury wastes into nearby waters. Dredging these deposits could free the mercury and allow it to enter the food chain.

The mercury scare prompted widespread research into the distribution of mercury in the environment and into the possible health hazards of eating fish with high levels of mercury in their tissues. Some microbes can convert relatively non-toxic mercuric ions into toxic methyl-mercury. These and other marine organisms concentrate poisonous methyl-mercury directly from the water or by eating contaminated organisms lower down on the food chain. Scientists have been studying the biological conversion of organic mercury to toxic forms, the biological concentration of mercury in the food chains, the occurrence of mercury in nature, and the toxicity of mercury in low levels to man.

One might assume that eutrophic lakes are likely to contain more poisonous mercury than younger oligotrophic lakes. But this would be wrong. D'Itri et al. (December 1971) found that fish from an eutrophic lake in Michigan contained less than half the concentration of methyl mercury (0.07 ppm) than fish from a younger oligotrophic lake nearby

The most common techniques for measuring mercury leave much to be desired since they usually identify only atomic mercury. Neither atomic absorption spectroscopy nor neutron activation analysis can distinguish between harmless elemental and mercurial compounds (such as mercuric sulphate), and highly toxic compounds (such as methyl mercury). Therefore, one must be wary of reports of mercury concentrations found in food or nature. It is wise to inquire how the determinations were made, and if they deal with toxic or non-toxic forms of mercury.

MERCURY LEVELS IN FISH AND OTHER FOODS

Are fish full of mercury and dangerous to eat? In general, the larger predatory fish have higher concentrations of persistent contaminants than do the smaller fish lower down on the food chain. But this has always been true. Despite industrial activity during the past 150 years, there appears to be little change in the mercury levels of fish in the sea. Isolated local changes yes, but no important increase in any species.

Scientists of the New York State Department of Environmental Conservation measured the mercury levels in fish caught in the late 1920's. They found levels of up to one ppm higher than the "safe" standards set by the FDA in 1970. Scientists at the New York Museum of Natural History found that tuna caught 90 years ago had mercury levels equal to or greater than those of contemporary fish.

Miller (1972) et al. studied the concentration of mercury in museum specimens of tuna and swordfish, and reported no significant change in mercury levels compared with contemporary fish.

An extensive survey of a non-problem posed by mercury in our food appeared in a report by J. T. Tanner (1972) and his associates at the FDA. The scientists tested between 20 and 35 samples each of food eaten by Americans such as flour, milk, sugar, potatoes, beef, shrimp, chicken, liver, and eggs. They found that all the samples were virtually free of mercury; most of the foods contained only trace amounts; and the highest single sample was found in a few shrimp which contained 43 parts per billion of mercury. (American authorities have rather arbitrarily declared that 5,000 parts per billion of mercury, when found in food, might pose a hazard to health). The Tanner group concluded that "With the exception of certain fish, the major foods in the United States are essentially free of mercury."

The levels at which mercury causes harm to humans are unclear. In 1967 the World Health Organization estimated that the concentration of mercury one could expect in one's diet due to natural causes and environmental contamination was 0.02 to 0.05 ppm mercury in its organic (methylated) forms. The FDA, when forced to establish a standard, chose a level 10 times more stringent, 0.5 ppm, as an "interim guideline." A Swedish panel of scientists has recommended that the 150 pound person ingest on the average no more than 0 03 milligrams of methyl mercury per day. A 200 gram (less than one-half pound) serving of fish with a 0.5 ppm level contains a total of 0.10 milligrams of methyl mercury, or 3 times the Swedish standard. Swedes who tend to eat a lot of fish have been advised by their government to eat less fish.

The FDA assumes that the average American eats fish less than twice a week, an assumption that helped determine the 0.5 ppm U.S. standard. The fish that caused Minimata disease in Japan had mercury concentrations roughly from 1.0 to 10 ppm, and were eaten daily by the people who fell ill. These conditions are quite rare in the world today.

Some individuals can tolerate much higher levels of a poison than others. Variations of over 10 to 1 in sensitivity to toxins are not uncommon

WHAT IS BEING DONE

Despite public fears generated by ignorance, it is apparent that no mercury poisoning epidemic exists in the United States, in fact there is no recorded case of any American falling ill due to mercury while on a normal diet. Most fish in U.S. waters contain less than 0.5 ppm mercury, though in a few areas the levels are greater than 1.0 ppm. Most other foods eaten by Americans contain less than 0.05 ppm, one tenth the recommended lower level. The Tanner study (1972) indicates that American food is relatively free of mercury, and no epidemics are likely to occur.

Vigorous measures have been taken to halt the discharge of mercury into the environment from known industrial sources. But there are nevertheless considerable amounts of mercury in rivers, lakes and oceans, and there always have been. Mercury is a long-lasting contaminant of the marine environment and is only taken out of the ecological cycle by incorporation in well-buried sediments. Stirring up of the sediments, as by dredging, may make the mercury available to the food chain.

Elimination of all large industrial point sources of mercury will not provide an absolute answer to the problem because relatively large quantities of mercury each year come from the improper disposal of manufactured items, such as thermometers and mercury switches; the burning of coal, the smelting of metallic ores, and from natural sources. Elimination of long-lasting mercurial fungicides, as has been done in Sweden, would reduce another major source of environmental mercury. In recent years, the United States has been using fewer mercury-based insecticides. Food is being monitored with increasing intensity for mercury and other poisons. Though measuring techniques are not always reliable or reproduceable, vigilance by government agencies keeps grossly polluted foods off the market.

LEAD IN THE ENVIRONMENT AND LEAD POISONING

"And for another 200 children there is no future at all, not even the vegetable-like existence of permanent institutionalization, for they will die as a result of this crippler of young children. Two hundred children a year."
Testimony of late Congressman William F. Ryan, March 9, 1972.

"It is estimated that each year lead poisons about 400,000 children of whom about 200 die."
Consumers Reports, November 1971.

To sum up the story of the 200 annual deaths from lead poisoning, Dr. Merlin K. DuVal, Assistant Secretary for Health and Scientific Affairs in the Department of Health, Education and Welfare, told a Senate subcommittee on March 10, 1972 that about 600,000 children are believed to have "significantly elevated lead levels...50,000 to 100,000 are likely to require treatment...lead paint poisoning results in 200 deaths annually..."

Who would not be unmoved by the deaths of 200 children each year? Not Congress. As a result of the fervent pleas of Congressman Ryan, who died of cancer in October

1972, Congress authorized $100,000,000 to fight the menace of lead in the environment. But let us see what lead is doing in our environment, how it effects human health, and then let us look for those 200 children who good authorities say are dying each year as a result of lead paint poisoning.

In the U.S. today, approximately one million tons of lead are used annually, most of it going for storage batteries and lead additives to gasoline. Use of lead in paint has diminished to where it accounts for only a minor percentage of total lead use, mainly in industrial anti-corrosion paints.

THE HAZARDS OF LEAD POISONING

The largest lead and zinc smelter in the world was closed by its management during March 1972 because of health hazards due to lead poisoning of workers. Imperial Smelting Corporation shut down its Avonmouth, England, plant, that was only four years old, after 24 workers became sick while working at the plant. More than 500 workers were found to have high levels of lead in their blood, that is, over 120 micrograms of lead per 100 milliliters of blood. Lead poisoning effects usually begin to show at about 80 micrograms per milliliter. None of the Avonmouth workers died, and most were able to return to their jobs after a few months of rest

Those who work with lead products have always been subject to lead poisoning, a hazard that can lead to disablement and death. It attacks workmen who fail to wash their hands after handling lead and thus transmit the particles to their mouth with food or tobacco. It attacks men who work in an atmosphere contaminated with metallic lead powders and aerosols such as are found in foundries and shipyards where ships are painted with red lead. Lead may enter the body through the alimentary canal or through the lungs. The body excretes most lead from the stomach but tends to retain lead in the lungs.

Petroleum products have a relatively low toxicity in themselves. But when gasoline is mixed with tetraethyl lead, it becomes quite dangerous if ingested. Tetraethyl lead is a notorious nerve poison, with toxic properties different from ordinary metallic lead poisoning.

Chronic lead poisoning results in "lead colic" in two out of three cases. The patient with lead colic suffers horrible abdominal pains, uncontrolled vomiting and acute constipation. Then lead paralysis, which often affects the fingers, may set in. The lead enters the blood stream and causes an anemic condition. Under normal conditions the body retains some lead in the bony skeleton, but most of it is excreted in urine. Severe lead poisoning will show up as a slate coloration of the teeth and nails. Lead poisoning results when high levels of lead circulate in the blood and enter the soft tissue. In its soluble form, lead can damage the kidneys and nervous system and even cause death.

Severe lead poisoning is often accompanied by convulsions, brain hemorrhaging and mental retardation. Many personality disorders and learning difficulties of slum children may be due to lead poisoning during a critical growth phase of their lives.

CHILDREN IN OLD HOUSES

For hundreds of years lead has been used extensively in paints, and lead pigments were used until the 1950's for most interior paints. After World War II, rural populations shifted to the cities and old dwellings began to deteriorate. Ignorance of the effects of lead poisoning have led to tragedy: as the lead-based paints peel off the walls of old slum buildings, young children often eat the chips. This tendency to eat non-food substances is called "pica" and is encouraged by the fact that lead paint has a sweetish taste. In the Middle Ages lead and lead acetate were used to "sweeten" wine.

Health authorities estimate that about 8 percent of city slum children suffered from some form of lead poisoning before the problem was noticed and corrective action was taken. Special credit goes to the Scientists Committee for Public Information (SCPI) scientists for investigating the lead problem and bringing it to the attention of the public. In cooperation with other community groups, the science information committees helped alert the community (Elwyn, 1968), informed the public, and were in large part responsible for the passage of federal laws in 1970 to deal with the problem.   Laws were passed to prevent the use of lead-based paints, and some money was made available for disease detection and paint removal. Attempts to persuade landlords to remove the toxic paints from their deteriorating houses are not always successful, since many landlords have no economic incentive to put money into old houses.   However, the programs of detection and amelioration appear to be working well.

New York City established a Lead Poisoning Control Bureau in 1969.   All hospitals in New York City have been called on to screen young children who may have been exposed to deteriorating housing.   In addition, 95 permanent facilities for testing are currently in operation, as well as mobile units and special door-to-door neighborhood programs.

When a child in New York City is found to have a lead level of 60 milligrams per milliliter or greater in his blood, the city Health Department notifies the agency or physician submitting the specimen.   A Health Department nurse and sanitarian visit the child's home.   The nurse discusses the situation with the family and helps them plan for medical care.   The sanitarian takes samples of paint and plaster from the apartment to determine sources of lead available to the child.   If the laboratory finds any paint samples with more than one percent lead content, the owner is ordered by the Health Commissioner to correct the condition within five days.   If the owner fails to comply the Emergency Repair Program of the city's Housing and Development Administration is requested to send a repair team to do the work.   When the work is done, the owner

is billed. To implement this program, in 1970 the Bureau of Lead Poisoning Control employed approximately 200 people and spent some $2,400,000 (Guinee, 1971).

Up to January 9, 1972, I was reasonably convinced that lead poisoning due to eating paint chips was a nasty business and that 200 children each year were dying from this particular environmental hazard. Then, by chance, I came across an article buried in the back pages of the New York Times. The article cited figures from the New York City Bureau of Lead Poisoning Control, which reported that in 1969, 10,000 children were examined and 727, or 7.3 percent of them were found to show signs of lead poisoning. In 1970, 80,000 examinations uncovered 2,255 cases of elevated leads in children tested.

During 1971, the Bureau tested over 115,000 children and found only 1,928 cases (1.8 percent) of elevated blood leads. It would appear that the poisoning rate has declined from over 7 percent in 1969 to less than 2 percent in 1971, and the number of deaths has declined from 12 in 1959 to 2 in 1969 to none in 1971. This heartening improvement shows what can be done when scientific knowledge is available and politicians are willing to cooperate.

Using the New York City data for 1971, I attempted to estimate the actual number of probable cases and deaths from lead poisoning throughout the country. My estimate is based on the following facts and assumptions:

About 70 million Americans live in large cities, and the 8 million residents of New York City represent some 10 percent of the U.S. urban population. However, many of the cities that have experienced fast growth since 1945, such as Los Angeles, Houston and Miami, do not have much in the way of dilapidated houses with lead-based paint. About 5 percent of New York City's housing units are substandard, according to Dr. Eli Ginsburg.

Since, in 1971, there were about 2,000 cases of elevated blood leads in New York City, then for the nation as a whole there should be no more than 20,000 cases, which is not quite as many as the 600,000 children figure cited by Dr. Merlin DuVal. As for deaths, from the figures given by the lead bureau in New York and from a separate analysis based on the fact that about 10 percent of the

United States population is between the ages of one and five, at which age "pica" is prevalent, and about 6 percent of all possible children have pica, I calculated that there should be no more than 10 deaths a year from lead poisoning, and not 200 as officially stated.

On March 11, 1972, I wrote to Dr. DuVal of the Department of Health, Education and Welfare, asking for facts regarding deaths. My letter was answered by the Director of The Bureau of Community Environmental Management, who cited National Bureau of Standards estimates of 2,500,000 children at risk. This figure is questionable, since there are only some 20 million children in the United States between the ages of one and five, and over 95 percent of them live in houses where there is no peeling lead paint, and therefore are not at risk. I wrote back asking Mr. Robert Novick, the Director of the Bureau, for data on the actual number of deaths. Eventually I received a curious reply from Novick saying that he would not send me the data I requested because it "would be subjected to misinterpretation." On August 24 I sent another letter to Novick, saying that if public officials were free to withhold public information merely because they deemed that it might be misinterpreted, then no citizen would ever find out what their public officials were up to. I again requested hard data on deaths from lead poisoning. No reply. On September 12, I called Robert Novick to find out why I had not received a reply. Mr. Richard E. Gallagher, Director of the Division of Community Injury Control, spoke with me and said that he thought a letter was on its way to me He noted that in 1968 the International Death Code had been changed to exclude deaths from lead poisoning, but doctors could note it on the death forms. However, he said that major cities, including Baltimore, Chicago, Philadelphia, Newark, and New York still report deaths due to lead poisoning. No letter arrived from Washington. I sent another query to Gallagher. On October 18 a letter with a report "Statistics and Epidemiology of Lead Poisoning (1972)" arrived. In this report I found the number of deaths in 1970 due to lead poisoning: Five!

LEAD IN THE AIR AND SEA

In the early 1920's, tetraethylead (TEL) was introduced as an anti-knock agent in gasoline. Deaths have been reported from drinking of leaded gasoline and by workers in TEL plants. In 1925 the U.S. Surgeon General appointed a committee to investigate the use and abuse of TEL. From this came a set of rules that were voluntarily adopted by the oil industry. This improved industrial workers' health, but did not prevent poisoning of those who drank leaded gasoline. In 1958 the Ethyl Corporation, the largest producer of TEL, increased the concentration of lead in gasoline from 2 to 4 grams per gallon of gasoline for use in high compression internal combustion engines.

Along heavily traveled roadways the concentration of lead on roadside vegetation and in the air is much greater than elsewhere. Cars discharge from 25 percent to 75 percent of the lead in TEL into the atmosphere, depending on driving conditions. Los Angeles, whose seven million inhabitants are habitually installed in its four million automobiles, has one of the highest average concentrations of airborne lead; five micrograms per cubic meter of air. The amount of lead in the world's atmosphere has jumped precipitously since 1940 and will continue to rise as car use increases, unless lead is removed from gasoline fuels.

Each year some 500,000 tons of lead in the form of minute particles find their way into the oceans of the northern hemisphere. Before industrialization, the rate of flow of lead into the oceans was some 10,000 tons per year. Since anti-knock gasoline came into widespread use, the rate has increased 50 times. Most of the lead falling into the ocean (270,000 tons) is washed from the atmosphere by rainfall, but over 200,000 tons enters the sea with water from rivers and runoff from the continents.

LEAD IN THE BODIES OF AMERICANS

We all ingest lead in our food. Every normal diet contains trace quantities of lead. These foods are not harmful as normally used. Adult Americans eat approximately 300 micrograms of lead per day, of which about 280 are excreted.

We also inhale airborne lead which lodges in our lungs. Lead in the air consists of particles in the micron (four-100,000th inch) range. Three-quarters of TEL-caused lead in the air has a diameter of less than one micron, a size that can easily be retained in the lungs. A rider on the San Diego Freeway, where the air often averages eight micrograms of lead per cubic meter during peak travel, will in eight hours breathe in about five cubic meters of air with 24 micrograms of lead in it. Most of this lead will be retained in the lungs. This is more than he retains from food eaten during a normal day.

Lead poisoning is a relatively insignificant preventable health problem now and should decrease even more in the future. Lead levels drop sharply once one gets away from streets with moving cars and trucks, and most Americans live in areas where the airborne lead level is well below the point of possible bodily damage.

Social pressure is forcing oil companies to remove the lead from gasoline. Automobile makers are willing to design engines that can run on unleaded gasoline, but the oil companies have balked at lead removal because it means expensive changes to refineries. But laws have been passed and lead is coming out of gasolines. Removing lead from gasoline has the added advantage that improved auto anti-pollution systems can be used without danger of lead fouling.

Implementation of the Occupational Safety and Health Act of 1970 and the Lead-Based Painting Poisoning Prevention Acts of 1970 and 1972 should reduce the dangers of childhood lead poisoning and occupational hazards even further.

Though lead poisoning is a horrible condition, it is a rare ailment in our society and should disappear completely in a few years. Both unions and management are cooperating to prevent lead sickness in industry; all indications point to a continued decline of childhood lead poisoning; and airborne lead caused by fuel additives is on its way to being eliminated. Since lead has never been shown to have beneficial effects on health, this reduction of lead in the environment can only be viewed with favor.

In answer to the basic questions posed on the environmental dangers of mercury and lead, after examining the available evidence, one can conclude that though lead and mercury have always existed in the environment, man has increased the level of lead in the air considerably since 1940. The overall increase of lead and mercury in the planetary biosphere is insignificant, however. Levels of mercury and lead in oceanic fish have not risen significantly during this century. While scientists disagree over the exact levels that should be prescribed as guidelines, the levels set by the FDA appear more than adequate to protect consumers. The rate of which both lead and mercury enters the environment is decreasing due to governmental and citizen concern. Compared with other causes of sickness and death, environmental poisoning by heavy metals is insignificant.

11

INSECTS, POLITICIANS, AND WEEDS

Human beings never welcome the news that
something they have long cherished is untrue;
they almost always reply to that news by
reviling its promulgators.

H. L. Mencken: Minority Report

Goodbye little bedbug, goodbye
Too bad you have to fly
Sad to see you go, Times have changed you know
Fates come between you and I.

Many was the night
When you and I held tight
Though we were close, don't be morose
Because of your upcoming flight.

I'm sure we'll meet again
In some secluded den
Until that day, with a twinge I say
Goodbye, darling bedbug till then.

Peter Agnos

BIOCIDES: DDT, IT'S HISTORY AND USE

Insects and spiders outnumber by far the humans on
this planet. These million or so species of arthropods
have, for eons, been waging a rather successful struggle
for existence. In the United States, some 10,000 insect
species are classified as pests. In addition, there are hosts
of weeds, fungi, nematodes (worms), rodents, and other
pests that have been competing with man for the natural
resources of the earth since humanoids first appeared on
the scene. Before the industrial revolution, mankind was
limited in the ways it could deal with its natural enemies.
Since the end of the 19th century, scientists have devised

chemicals to combat man's biological enemies. A few of these biocides have proved singularly effective. However, some of the chemicals have real or imagined harmful side effects. Among these latter compounds are the important insecticide DDT, and the herbicide 2, 4, 5-T.

DDT, a synthetic organic insecticide, was developed in Germany in 1874. It was first widely used to combat typhus-spreading lice during the Second World War. DDT retains its toxicity to insects over a period of months and sometimes years. A prime advantage of DDT and other "hard" organic pesticides is precisely their tendency not to biodegrade readily and hence their ability to kill insects over a relatively long period of time without repeated application. As a result of the widespread use of DDT and other related chemicals such as dieldrin, farmers throughout the world have been able to increase food production, and some of the most virulent scourges of mankind, such as malaria, typhus, and encephalitis, have been virtually wiped out over large sections of the earth.

Even before the advent of DDT, persistent pesticides and poisons were used to combat pests. Arsenic, thallium, and sulfur compounds have been sprayed or dusted on crops; petroleum derivatives and pyrethrins have been used against flies, mosquitoes, bedbugs, roaches, and other cosmopolitan pests. Arsenic, thallium, mercury, and other pesticide compounds like DDT, are not readily biodegradable and can become incorporated into the food chain of animals and man.

DDT is soluble in fats and thus may be cycled through ecological food chains, going, for example, from a plant to a grasshopper to a robin to a hawk and increasing in concentration in each higher predator. The levels of DDT tend to rise in animal tissue, especially fatty tissues, as one predator feeds on other animals lower down in the ecological cycle, so that the human American omnivore carried in 1970 about 12 parts per million of DDT in his fatty tissue (the Council on Environmental Quality's 1971 report gives 5 ppm for children and slightly more for adults as a whole). It is doubtful if this concentration is harmful to humans in any way. However, concentrated doses of DDT can harm insects other than the ones regarded as pests, and also adversely effect the behavior and reproduction of a few species of birds and small animals.

Other insecticides related to DDT are aldrin, dieldrin, chlordane, endrin, heptachlor, and toxaphene. Less persistent but usually more toxic substitutes for DDT-like insecticides include the organophosphates, parathion and malathion, and the carbamates, which were responsible for approximately 200 deaths and thousands of severe illnesses in the United States in 1970. Wildlife has also inadvertently been destroyed through misuse of these newer substances.

Since the publication of Silent Spring in 1962, the use of DDT has been declining in the United States, though it is being widely used in other countries to control malaria and increase food production. DDT has saved at least five million lives and untold millions from disease and sickness and has generally increased the food available to mankind. Modern America could not feed itself adequately, to say nothing of having a food surplus, without the use of insecticides. Man has established his own natural balance on the great mechanized farms throughout the Soviet and capitalist world. Today most large farms are chemical factories maintained by chemical weed killers, fertilizers, and insecticides, and powered by irrigation pumps and agricultural machinery. Nevertheless, despite the American farmer's sophisticated arsenal of insecticides, insects consume and destroy $3 billion to $20 billion worth of crops each year. The National Agricultural Chemical Association says "without pesticides, crop and livestock output would be reduced by about 40 percent. Farm exports would be wiped out and the price of food would increase anywhere from 50 percent to 75 percent."

Dr. Norman E. Borlaug, Nobel Peace Prize winner in 1970, is the father of the postwar "green revolution," which gives the only hope for feeding an expanding world population. In an address before the 16th governing conference of the United Nations Food and Agriculture Organization (FAO) in Rome on November 8, 1971, Dr. Borlaug said that if world agriculture is denied the use of agricultural chemicals "because of unwise legislation now being promoted by a powerful group of hysterical lobbyists, who are provoking fear by predicting the doom for the world through chemical poisoning, then the world will be doomed not by chemical poisoning, but by starvation."

Dr. Borlaug's position is similar to that of the FAO. The FAO has said repeatedly that until cheap, safe and efficient substitute pesticides are available, "there is no alternative to the judicious use of DDT," especially in the developing nations.

Answering Dr. Borlaug in the New York Times (December 7, 1971), Dr. Barry Commoner, the environmentalist, said "I am unable to accept Dr. Borlaug's claim that a severe reduction in the use of agricultural chemicals would sharply reduce agricultural output." To back his argument Commoner cited a recent symposium where it was suggested that the U.S. could accomplish a 70 percent to 80 percent reduction in insecticide use with no reduction in output "simply by increasing harvested acreage by 12 percent." Dr. Commoner, a biologist, may not be aware of the fact that about a million acres a year are made unfit for agricultural use by road building, expansion of industry and urban areas and acquisition for recreational use; and that an agricultural land expansion of 12 percent would be far from simple.

In his Rome talk, Dr. Borlaug took particular exception to Rachel Carson's book, Silent Spring, that he said had precipitated "the vicious hysterical propaganda campaign against the use of agricultural chemicals. It was a diabolic, vitriolic, bitter one-sided attack on the use of pesticides..." in which DDT had been made the main villain. Let us look closer at Silent Spring.

SILENT SPRING AND AFTER

Few books have had as great an impact on environmental thinking as Silent Spring, which was published in 1962. In her angry attack on the use of pesticides, Rachel Carson succeeded in frightening millions of people. It is an important book, although it is hard to go along with Justice William O. Douglas, who has called it "the most important book of the century."

According to Miss Carson, we are so surrounded by pesticides that "we are in little better position than the guests of the Borgias." She wrote that "one in every four" of us will die of cancer as a result of ingesting cancer-causing chemicals in our food. According to Miss Carson, chemicals are destroying wildlife to the extent that robins are on "the verge of extermination," and thus with the birds gone, we will shortly experience "silent springs."

In Silent Spring Miss Carson apparently let her concern for animals affect her regard for reality. To cite one flagrant example, far from being on the "verge of extinction," the robin, according to the Audubon Society, increased over a thousand-fold from 1941 through 1960 and is now among the most numerous birds in North America. The National Academy of Sciences reported in 1963 that many statements in Silent Spring were misleading, and concluded that "pest control to protect human health, food, fiber and forest and other biological resources is essential by whatever means necessary."

Still the Carsonites persisted in their criticism of DDT, and in 1966 a Senate subcommittee headed by Senator Abraham Ribicoff studied the matter. The subcommittee found and stated that "no significant health hazard exists today (from DDT)... It has been fed to human volunteers in significant doses for 18 months without demonstrable acute effects." The tests referred to were extended and well-controlled tests conducted by the U.S. Public Health Service, using volunteer prisoners to see if unusually massive doses of DDT were harmful to humans. All indications were that DDT is not toxic to man.

DDT AND WILDLIFE

Among the "hysterical lobbyists" cited by Dr. Borlaug as "the moving forces behind the environmentalist movement" seeking to ban DDT are the Sierra Club, the National Audubon Society, the Isaak Walton League and the Environmental Defense Fund. The first three groups have long had an interest in wildlife and wilderness.

What effect, if any, has DDT had on wild animals, fish and trees? We have already noted the increase in the robin population, despite Miss Carson's opinion that these poor birds were on the "verge of extinction." As for other birds, some appear actually to benefit from ingestion of

185

DDT. In a study reported in Nature in 1969, D. J. Jeffries reported that Bengalese finches developed thickened egg-shells when fed DDT. In other bird studies, R. G. Heath (Nature, 1969) found that DDT appeared to aid reproduction in mallards. I cannot explain these findings; I merely cite them.

A few American bird species that prey on fish are declining in numbers. These birds include brown pelicans, bald eagles, ospreys, peregrine, falcons, and glossy ibises. The cause of the decline, which usually manifests itself in thin eggshells, is still in question. Animals in the wild may be subject to a host of insults and stresses from egg collectors, hunters, jet plane noises or from a variety of deadly and long-lasting poisons, any of which could account for thinning of eggshells. Consider the case of the declining bald eagle. Since 1964 these majestic birds, when found dying or dead, have been frozen and sent to the U.S. Fish and Wildlife Service Laboratory in Laurel, Maryland, where the chemicals in their tissues are analyzed. Of 34 birds recently examined with sufficient chemicals in them to kill, 22 contained toxic amounts of thallium, two of mercury, one of lead, and eight of dieldrin. But only one out of 34 contained excessive DDT. These five toxic chemicals are persistent; all adversely affect reproduction in birds. For certain birds an eggshell-thinning correlation with DDT dosage has been demonstrated in the laboratory and found in nature. But it would be strange if DDT were the only cause of eggshell thinning, since DDT has been declining in use for the last 12 years, while thin eggshells have only become a noticable problem in the past few years.

While some of the birds of prey mentioned above are declining in some locations, there is little chance that they will become extinct. At the end of 1969 some 400 bald eagles were sighted in Glacier National Park, a record number. The bald eagle has also been reported on the increase in other areas, according to the Audubon Society. Declines in particular areas such as the western range and Alaska are almost certainly due to hunters who stalk the big birds with guns, planes, and poisoned bait. Well over 70 percent of all American eagles die unnatural deaths due to bullets and poison.

The peregrine falcon is reported to be having thin-eggshell problems in Wisconsin, where the average falcon has about one ppm of DDT in its fatty tissue. However, in Canada, where the concentration in the fat of peregrines is much greater than in Wisconsin, there appears to be no reproductive problem. Ospreys appear to be declining in numbers in the DDT-less Long Island Sound area, but they are doing quite well in the Chesapeake Bay area. Vocal anti-DDT factions on Long Island have forced local authorities to stop using DDT in mosquito control programs since 1966. One possible cause of birth defects and declines in the ospreys of Long Island could be the high-intensity radiation experiments being performed by the Brookhaven National Laboratory in an open wooded region of eastern Long Island. Any osprey happening to fly close to the powerful radiation source could very easily suffer genetic damage

In general, America's birds are doing well. In California, the Sierra Club's home state, licensed hunters annually bag about 2.5 million quail, 750,000 pheasants, 350,000 geese and 2.5 million ducks, among other numerous birds.

Fish, according to some environmentalists, are the victims of DDT. In 1963-1964, a massive fish kill in the lower Mississippi River was blamed on endrin, a DDT-like insecticide manufactured in a Memphis, Tennessee, plant. However, the fish kill occurred about 50 miles downstream from the plant, while the fish in the intervening length of the river were unaffected, making it virtually certain that the Memphis plant was not to blame. Fish die-offs occur in nature due to diseases, or the growth of poisonous plankton such as redtide dinoflagellates. Massive animal die-offs are a natural phenomena when the environment changes.

C. F. Wurster, in a paper published in Science in 1968, implied that the planetary oxygen supply was in danger because DDT was killing off marine algae, small floating plants that (like land plants) convert carbon dioxide to oxygen. Wurster, the chief scientist of the Environmental Defense Fund, reported that in a laboratory experiment he had found that 10 parts per billion (ppb) of DDT in water seriously interferred with photosynthesis in five species of phytoplankton. "Algae," warned Wurster, "are responsible for more than half the world's photosynthesis; interference with this process could have profound world-wide biological implications."

In a talk given at Yale as part of a symposium on "Issues in the Environmental Crisis" (1970), Wurster elaborated on his research and its implications as follows:

A few years ago I did some studies at the Woods Hole Oceanographic Institution in which the photosynthesis of marine phytoplankton was measured where DDT had been added to the water in the parts-per-billion range. The data indicated that as DDT was added to the water, the rate of photosynthesis was decreased. (Photosynthesis is the process whereby green plants absorb carbon dioxide and the energy from sunlight, producing organic nutrients and oxygen. All animal life on earth is dependent on this process.) By the time we reach eight or ten parts per billion of DDT in the water, the photosynthetic rate is diminished to about three-quarters of normal; as the DDT concentration goes higher, the photosynthetic rate goes lower. Since these experiments were done, others who have worked in the field have confirmed the findings in both marine and freshwater systems.

What are the environmental implications of these results? (I find it difficult to extrapolate from an Erlenmeyer flask to an ocean). There is no evidence that DDT is reducing the rate of the photosynthesis in the ocean, nor that DDT is present there in comparable concentrations. In local areas, I think it means that if DDT reaches these concentration ranges, it will inhibit the whole base of the food chain. I believe it also indicates a likelihood of manipulating the species composition of the phytoplankton community because it is selectively toxic. Since it does not exert the same toxic stress on all species, it could aggravate problems of eutrophication (overenrichment that leads to algal blooms). It could increase algal-bloom problems where certain species that are not good food organisms

188

could become predominant. If this should occur, the species composition of the whole food chain might be changed.

What Wurster did not say is that the average concentration in the oceans of DDT and its metabolites (DDE, DDD) is at most one part per trillion (ppt), one thousandth that of the no-effect level. It is therefore virtually impossible for the DDT levels (cited by Wurster as harmful to some plankton) ever to be reached in the oceans. In the MIT report, "Man's Impact On The Global Environment," a panel of scientists said:

> The effect of DDT on the ability of the ocean phytoplankton to convert carbon dioxide into oxygen is not considered significant. The DDT concentrations necessary to induce significant inhibition exceeds expected concentrations in the open oceans by ten times its solubility in water.

What about shellfish? In an attempt to find out how oysters react to high levels of DDT, Environmental Protection Agency scientists have been exposing oysters to flowing seawater with one to three ppb of DDT, parathion, and toxaphene for over two years at the Government's Gulf Breeze, Florida, laboratory. Growth and development of the oysters were recorded at two-week intervals. Oysters exposed only to DDT for 12 weeks acquired a residue to 76 parts per million, which is 76,000 times as great as the concentration in the flowing seawater. The oysters did not die, and when returned to ordinary seawater, purged themselves of almost the entire accumulated DDT residue.

In the words of the Fish and Wildlife Service report:

> When oysters were reared from juveniles to sexual maturity in flowing seawater chronically polluted with low levels (three ppb or less) of DDT, toxaphene and parathion...over a period of two years, the loss of weight was about 10 percent of total body weight. Weights and heights of separate groups of oysters

reared in seawater containing about one ppb of either DDT, toxaphene or parathion were not statistically different from controls... Eggs and spermatozoa removed from the oysters developed into 24-hour trochophore (newly hatched) larvae. The oysters accumulated relatively high levels of DDT and toxaphene but eliminated them during a three month depuration period.

Commenting on the level of DDT in oysters and in the estuaries where oysters live, Dr. Donald Spencer said in 1971:

> The National Estuarian Monitoring program (formerly Bureau of Commercial Fisheries, now EPA) has for better than four years collected oysters (or certain other molluscs) at monthly intervals from some 165 sites around the coast of the United States. Of the nearly 7,000 analyses performed, approximately 98 percent showed residues of DDT plus its metabolites of less than 0.5 ppm. The residues fluctuated sharply during the year in oysters from the same site reflecting use patterns of DDT in the drainage basin. They appeared to lose residues almost as rapidly as they acquired them, often dropping back to levels approaching the sensitivity of the analytical method.
>
> If under controlled conditions an oyster exposed to 1.0 ppb for 12 weeks acquires a residue of 75 ppm, then oysters acquiring only 0.5 ppm in the natural environment (and this only at times when new materials are being used in the drainage basin) must indicate that a very small percent indeed of DDT must be reaching the estuaries. In this connection it is of interest to record that the residue levels in oysters were so low off the shores of Washington, Maine, and South Carolina that the cooperators in these states elected to drop the program.

As for other wildlife, one would expect large game and birds to decline as man expands his activities and forest cover declines. But curiously enough there is more game in America today than there was 30 years ago! In the eleven western states in continental North America the annual harvest of elk and deer has more than doubled in the past 20 years. The antelope population has grown from a few specimens around 1900 to a yearly harvest of 25,000 animals in Wyoming alone. Beavers, which were trapped to near extinction for their pelts, have become something of a pest in the West because they gnaw down trees and flood small rivers. In addition, non-native species such as wild pigs, Barbary sheep, turkeys and horses run wild through many western states.

The great increase in wildlife in the United States has come about because, since we need less land for growing crops, some 80 million acres have been taken out of production in the past 20 years. During the past 30 years, we have learned a lot about managing natural wildlife resources. The U.S. government has set aside over 100 million acres as natural parks and preserves. In addition, the Sierra Club, the Nature Conservancy and private individuals control and manage many hundreds of thousands of other natural preserves, and farmers care for additional millions of acres of woodlots in which birds and animals thrive.

However, our forests are not invulnerable. Lately the gypsy moth and other insect pests have attacked. Because of Silent Spring and its unsilent offspring, chemical control methods are politically unpopular, and many states have prohibited or drastically controlled the use of DDT-like substances. The gypsy moth was introduced into Massachusetts in 1869 for experimental purposes. It escaped and has spread throughout the Northeast. In 1971 the pest was reported as far south as the Carolinas and it is spreading with alarming rapidity. In Connecticut the acreage under attack increased from 369,000 acres in 1970 to over 655,000 in 1971. In New Jersey the increase is from 10,000 acres a few years ago to over 190,000 in 1971. Some 15 million acres were denuded in the Northeast

191

in 1971 and more than that was stripped in 1972 because pesticides were not used. But chemicals such as DDT are in disfavor, and no good substitute exists.

BIOLOGICAL INSECTICIDES: ALTERNATIVES TO DDT?
Let us use natural techniques to combat the thousands of insect pests, suggest the children of Silent Spring. It would be beautiful if we could.

Among the suggested biological anti-insect techniques are the following:

1. Radioactive substances or chemicals that would sterilize insects. In the Knipling technique males are sterilized in large numbers and then allowed to mate with females which then produce no offspring. Screw worm flies and Florida fruit flies have been eradicated from selected areas using this method. It is particularly effective on small, isolated islands. Its cost is quite high.

2. Sex attractants such as dispalure mimic the scent with which female moths attract males. Aerial spraying of a wood with dispalure makes it difficult for the males to find a mate. This technique has been used experimentally against moths with some success, but again at high cost.

3. Some insect parasites attack specific insect pests. For many years ladybug beetles have been raised to eat aphids; ichneumon wasps deposit their eggs in other insects upon which the wasp larvae feed; and praying mantis eat other insects.

4. Viruses and bacteria attack various specific insects.

5. Birds and other small animals feed on insects.

In the New York Times of August 22, 1971, C. F. Wurster wrote that "we have had numerous safe and effective insecticides for years." He implied that he knew of "superior alternative methods for controlling insects." As of today, though biological controls have proved effective only in certain localities against only a few selected insects; they are still merely promising and uneconomical where available. Dr. E. F. Knipling, a leading exponent of biological controls who developed the male sterilization technique, said (1971) that "We still need to rely on proved pesticides. The alternatives are going to come slow and will take time. But in the end I'm hoping that we can truly manage a number of key insects effectively and economically." Now we have only chemical controls.

DDT AND CANCER

More people are dying of cancer, and well they might. Since fewer humans are now dying of tuberculosis, malaria, typhus and other infectious and insect-borne diseases, they must nevertheless die of some cause eventually, and hence degenerative diseases such as cancer and cardiovascular failure are on the increase. As cancer deaths increase in frequency, more people grow to fear it. Hence any imputation that substance X causes cancer is enough to frighten millions of innocent souls into demanding its ban. It is noteworthy that the life expectancy of Americans in 1970 is 20 years greater at birth than it was in 1900 and that all age groups of Americans are living longer now than they did in "the good old days."

The opponents of DDT have opposed it on the grounds (among others) that it may cause cancer. Wurster in his Times article (August 22, 1971) said, "The insecticides DDT, aldrin, dieldrin, heptachlor, and mirex (all long-lived chlorinated hydrocarbons) have been shown to cause cancer in laboratory animals. Such tests do not prove conclusively that these materials cause cancer in man, but they indicate a strong probability that they are human carcinogens." Other more rabid environmentalists have gone further than Wurster in their statements.

But the tests cited by Wurster and others usually involve large amounts of substances injected under the skin of laboratory animals such as mice or rabbits. Using similar techniques, it has been shown that honey, charcoal-broiled steak, and raisins are also carcinogenic. It is simply impossible to jump from these animal experiments to any realistic conclusion about the effects of DDT on man. What we do know is that volunteers ingested large quantities of DDT for two years and showed no ill effects.

Dr Edward R. Laws, Jr. of Johns Hopkins Hospital reported on a study that adds experimental support to an observation made among workers exposed on their jobs to high levels of DDT for 10 to 20 years. According to

Laws, "It is noteworthy that no cases of cancer developed among these workers in some 1,300 man-years of exposure, a statistically improbable event."

Eighty-nine DDT-fed mice were innoculated with tumor cells: only seven did not develop a cancer. However, all 87 mice without DDT developed cancer. All cancerous mice died. Those with DDT in their systems lived significantly longer!

One of the most powerful arguments against the widespread use of DDT has been the possible cancer producing effect on man, arguments based on experimental development of liver cancers in mice given extraordinary high doses of DDT for long periods of time.

"Evidence which is at least as impressive," Laws said, "leads to the opposite conclusion; namely, that DDT may have an anti-cancer producing potential. With a technique that produces tumors 100 percent of the time, it is impressive that seven animals on DDT never developed tumors."

In 1969 the Audubon Society distributed widely about 700,000 copies of an anti-DDT broadside. The leaflet concluded, "DDT should be banned throughout the land, and banned for export. We welcome your support in this campaign." The campaign has been relatively successful in that many states have banned DDT. A number of European countries jumped on the bandwagon. Ceylon was one of the first Asiatic countries to ban DDT, with startling results. More than two million Ceylonese had malaria in the early 1950's when DDT was first introduced to control malarial mosquitoes. After 10 years of control, malaria had all but been eliminated in Ceylon. The country banned the pesticide in 1964. By 1968 over a million new cases of malaria had appeared. Ceylon rescinded its ban on DDT in 1969. Another result of the anti-DDT crusade is that the amount of DDT in use in the U.S. has declined by more than 50 percent in the last ten years. Only one company, the Montrose Chemical Company, continues to make DDT, and most of it is for export.

The anti-DDT forces persisted despite feelings of the World Health Organization that "DDT has been the main agent in eradicating malaria in countries whose population totals 550 million people; of having saved about five million lives and prevented 100 million illnesses in the first eight years of its use; of having recently reduced the annual malarial death rate in India from 750,000 to 1,500; and of having served at least two billion people in the world without causing the loss of a single life by poisoning from DDT alone..." The average lifespan in India increased from 32 years to over 47 years as a direct result of DDT use.

In the late 1970's, several groups led by the Sierra Club, the Audubon Society and the Environmental Defense Fund went to court in an attempt to force the government to ban DDT. In January 1971, in a two-to-one ruling, the U.S. Court of Appeals in Washington, D.C., ordered the federal government to issue immediate notices of cancellation of all uses of DDT and to determine whether DDT was an "imminent hazard" to public health. William Ruckelshaus, administrator of the EPA, was directed to carry out the order. The Department of Agriculture had already banned the use of persistent pesticides on many food crops.

The issue is still before the courts and the administration in Washington. It was under this circumstance that Dr. Norman E. Borlaug gave his passionate speech in Rome pleading for the continuance of DDT use. Following are excerpts from his speech:

> The current vicious, hysterical propaganda campaign against the use of agricultural chemicals, being promoted today by fear-provoking, irresponsible environmentalists, had its genesis in the best-selling, half-science-half-fiction novel, Silent Spring, published in 1962. This poignant, powerful book, written by the talented scientist Rachel Carson, sowed the seeds for the propaganda whirlwind and the press, radio and television circuses that are being sponsored in the name of conservation today (which are to the detriment of world society) by the various organizations making up the environmentalist movement.
>
> The environmentalists would like to have a legislative ban placed on DDT so as to prohibit it for any use in the United States. Almost certainly as soon as

195

this is achieved, these organizations will begin a world-wide propaganda barrage to have it banned everywhere. This must not be permitted to happen, until an even more effective and safer insecticide is available, for no chemical has ever done as much as DDT to improve the health, economic and social benefits of the people of the developing nations.

Although more than 1,400 chemicals have been tested by WHO for use in malarial campaigns, only two have shown promise and both of these are far inferior to DDT.

It is now obvious that the current aim of the Environmental Defensive Fund and its affiliated environmentalist lobby groups is to ban DDT first in the U.S. and then in the world if possible. DDT is only the first of the dominoes. But it is the toughest of all to knock out because of its excellent known contributions and safety record. As soon as DDT is successfully banned, there will be a push for the banning of all chlorinated hydrocarbons, then in order, the organic phosphates and carbamate insecticides. Once the task is finished on insecticides, they will attack the wood killers and eventually the fungicides.

If the use of pesticides in the U.S. were to be completely banned, crop losses would probably soar to 50 percent, and food prices would increase fourfold to fivefold. Who then would provide for the food needs of the low income groups? Certainly not the privileged environmentalists."

Dr. Charles Wurster of the Environmental Defense Fund replied to Dr. Borlaug's attack by saying Borlaug was "one or two decades out of date." Dr. Elvis Stahr, President of the Audubon Society, said that environmentalists were not being merely negative but were seeking safe controls of pests. A Sierra Club director, William Futrel, wrote a letter to The New York Times suggesting that Dr. Borlaug was connected with the pesticide industry. The fate of DDT for use in today's and tomorrow's world is still undecided.

POSTCRIPT ON MEASURING DDT

Measurements around the world appear to indicate increasing amounts of DDT and its metabolites in the ocean and in the soil. But is it DDT that is being measured? Three scientists, Frazier, Chesters, and Lee, of the University of Wisconsin's College of Agriculture, tested samples of preserved soils for DDT. In 1970 they reported that they had found "apparent" organochlorine insecticides in soils collected in 1910! Of the 34 samples studied, 32 showed apparent DDT-like insecticide residues.

This is strange since the DDT family of insecticides was not widely used in the environment until the 1940's. Professor G. B. Lee wrote me (letter of January 12, 1972) that "At the present time we do not know what substances in the soil produce peaks similar to certain pesticides. They do, however, appear to be indigenous soil compounds." One may well wonder if much of the fear of DDT accumulation in the biosphere is not based on misleading data; that is, measurement of natural compounds.

PCB's

PCB's (polychlorinated biphenyls) are colorless, odorless industrial liquids that resemble DDT in that they biodegrade slowly. They are used primarily as insulators in high voltage equipment and are manufactured by only one company in the United States, Monsanto Chemical Company, and in a few other factories abroad. PCB's can withstand temperatures of up to 1,600 degrees Fahrenheit, which makes them useful in equipment where heat must be transferred. They are also used as additives in certain paints, copying papers, plastics, and insecticides. As industrial chemicals go, they are not widely used.

PCB's have been found throughout the world in animals, water bodies and in man. They enter the environment accidentally when, for example, a large power transformer leaks, or when plastic or cardboard contaminated with PCB's come into contact with food; or when substances containing PCB's weather or wear away and the long-lasting PCB molecules wander into the biosphere. The Environmental Protection Agency has found some human tissue with concentrations as high as 250 parts per million (ppm). It is not likely that anyone has died from poisoning by the PCB's, but because of their persistence and similarity to DDT they have been vigorously attacked by environmentalists. Scientists employed by the government who have looked into the PCB matter have stated that as they are presently used, PCB's cannot enter the food chain or the environment in amounts that will cause damage.

Nevertheless, over 125,000 fowl were destroyed during the summer of 1971 when PCB's were detected in fish meal, chicken eggs, and turkeys. Apparently a power transformer had sprung a leak in a factory making fish meal and some of the PCB's being used in insulators had leaked into the meal. The meal was fed to chickens and turkeys as a feed supplement, and the PCB's were inadvertently propagated down the food chain. When the food containing PCB's was discovered, it was destroyed. Perhaps the fowl were then given a special burial in lead lined caskets to prevent further circulation of the PCB's in the biosphere.

At least twice during 1971 The New York Times reported that three persons had died in Japan in 1968 as a result of eating food fried in cooking oil contaminated with PCB's. I could find no corroboration for the three deaths and consequently wrote to the Times asking for references on this matter. To date no reply has been received. A. T. Hammond (Science, January 14, 1972) reported, "It is not clear whether any of the subsequent deaths among the patients (sickened by eating the contaminated oil) can be attributed to acute PCB poisoning, and it is still not known whether the commercial PCB mixture involved contained traces of dibenzofurans or other impurities."

There is no direct evidence that PCB molecules per se, which are relatively inert, can cause any harm to man. The Monsanto Corporation has been conducting tests for the past two years with animals and has found no harmful effects. Nevertheless, such is the climate regarding strange chemicals today that the U.S. Food and Drug Administration placed a limit for PCB's of five ppm on chickens and 0.5 ppm for eggs. This level is arbitrary. There is considerable doubt whether it is PCB's or another chemical, chlorinated dibenzofuran, sometimes found with PCB's, which may be harmful. The harm found to date involves laboratory animals who, when subjected to large doses of PCB (and chlorinated dibenzofuran?), appeared to suffer birth defects.

Though the danger from PCB's diminishes due to strict controls imposed by the only manufacturer in this country, the issue still exercises some fertile imaginations. A group of investigators published an article (Mosser, et al., 1972) in Science describing the effects of dousing five species of algae with PCB's. The article concludes with the following paragraph:

> Selective inhibition of sensitive phytoplankton species by PCB's, DDT and other stable pollutants in the environment may alter the species composition of natural algal communities. Such effects at the base of aquatic or estuarine food webs could profoundly affect higher organisms as well.

It is simple to assert that an event, having once occurred in a laboratory, "may" take place again. Thus it took no great foresight for Mosser et al. in their study of PCB effects to conclude as they did. The authors' conclusion that selective species alteration "may" occur is indisputable but obvious, since the addition to an ecological system of almost any long-lived contaminant is likely to alter species composition.

An estimate of the probability of ecological damage likely to occur due to PCB usage would have been meaningful. C. Gustafson (1970) has pointed out that "All studies of PCB's in animals indicate that acute toxicity is not a significant factor...," and further, "Compared to DDT...PCB's have a relatively low acute toxicity..." That PCB's "could profoundly affect higher organisms" is dubious because, as the authors failed to mention, PCB distribution is much more tightly regulated than it was a few years ago, and the levels prevalent in the environment are likely to cause little, if any, harm to plants or to animals. Statements to the contrary sound like the wishy-washy fantasies that warned mankind that "robins were on the verge of extinction" and the supply of planetary oxygen could diminish due to DDT usage.

2, 4, 5-T BANNED: THE DECISION AGAINST REASON

"The question which has been raised recently concerning the hazards of 2, 4, 5-T and related chemicals may in the end appear to be much ado about very little indeed. On the other hand they may ultimately be regarded as portending the most horrible tragedy ever known to mankind."

Senator Philip A. Hart, April 1970

The herbicide 2, 4, 5-T is one of the most important chemicals developed since World War II to increase the world's food supply and to clear waterways of weeds, the remnants of wild vegetation. It is one of a group of herbicides which cause broad-leaved plants to grow so rapidly that they cannot sustain themselves and hence die out, leaving the ground barren so that food plants can be grown. Georg Borgstrom (1971) wrote:

Canada's grain-growing states, Manitoba, Saskatchewan, etc., expect average annual losses ascribed to weeds of 10 to 15 percent; England from 7 to 10 percent of its grain production; India 20 to 30 percent. If the weeds could be eliminated, the Indian subcontinent would under prevailing growing conditions obtain a 25 percent larger harvest. Still more surprising perhaps is that the Unied States, despite full use of modern chemical weapons, suffers an annual loss valued at four to five billion dollars.

Weeds compete with crops for space, light, nutrients, and water.

Substances such as 2, 4, 5-T have been used devastatingly in Vietnam by the U.S. Army and South Vietnam to destroy crops and forest cover that could be used by the Viet Cong.

In late 1969, President Nixon's Science Advisor, Dr. Lee DuBridge, announced that actions were being taken to restrict the use of 2, 4, 5-T in the U.S. This action was precipitated by laboratory findings that some rats and mice who were fed large doses of the herbicide during early pregnancy gave birth to defective offspring.

The announcement, together with reports by South Vietnamese newspapers of an increased occurrence of birth defects during June and July 1969, elicited far-reaching reactions from governmental agencies. A few scientists, environmental groups, and public media called for outright bans. Government-sponsored panels of experts, special commissions set up by scientific organizations, hearings before subcommittees of the U.S. Senate, and conferences attended by representatives from industry, government, and universities examined available data and heard expert opinions. None of these groups, however, was able to provide a generally acceptable answer to the central question of whether 2, 4, 5-T, as then currently produced and used, constituted a risk to human pregnancy or to health.

In April 1970, the Secretary of Health, Education and Welfare advised the Secretary of Agriculture that "In spite of these uncertainties, the Surgeon General feels that a prudent course of action must be based on the decision that exposure to this herbicide may present an imminent hazard to women of child-bearing age." Accordingly, on the following day the Secretaries of Agriculture, of Health, Education, and Welfare and of the Interior jointly announced the suspension of the registration of 2, 4, 5-T for: "I. All uses in lakes, ponds or in ditch banks. II. Liquid formulations for use around the home, recreation areas and similar sites." A notice for cancellation of registration was issued on May 1 for: "I. All granular 2, 4, 5-T formulations for use around the home, recreation areas and similar sites. II. All 2, 4, 5-T uses on crops intended for human consumption." Producers of 2, 4, 5-T were advised of these actions, and two of the registrants, Dow Chemical Company and Hercules Incorporated, exercised their right under the Federal Insecticide, Fungicide and Rodenticide Act to petition for referral of the matter to a Federal Advisory Committee.

The National Academy of Sciences set up an advisory committee that was to (1) consider all relevant facts; (2) submit a report and recommendations regarding registration for certain uses of 2, 4, 5-T; and (3) state the reasons or bases for these recommendations. The committee's main task was to determine whether the use of the herbicide did in fact constitute an imminent health hazard, especially with respect to human reproduction. Accordingly, the committee examined all available information and evaluated its relevance to the potential hazard of human exposure during pregnancy.

In judging the effects of 2, 4, 5-T, the prestigious committee considered appropriate issues as follows:

1) As is frequently the case, available data are insufficient for a definitive statement of conditions under which a specified risk might occur, assuming that freedom from risk is ever attainable.

2) Since most chemicals under suitable laboratory conditions could probably be demonstrated to have teratogenic effects, and certainly all could be shown to produce some toxic effects if dosage were raised high enough, it would not be reasonable to consider the demonstration of toxic effects under conditions of greatly elevated dosage sufficient grounds for prohibiting further use of a particular chemical.

3) Benefits are to be expected from the continued use of 2, 4, 5-T. The necessity of making a value judgment of benefit vs. risk, therefore, must be accepted, not only for this herbicide, but for numerous valuable drugs, some natural nutrients, and many other chemicals, some of which are known to be teratogenic in laboratory animals. The risk vs. benefit judgment for a particular herbicide or drug can be evaded only if it can be shown that another compound is equally as efficient and involves less risk. This presupposes that the risk potential of a substitute herbicide is at least as well known as that of the original (in this case 2, 4, 5-T), a fact that may be difficult or impossible to ascertain. The substitution of a relatively unknown pesticide for an older one with known adverse effects

is not a step to be taken lightly. Even with steadily improving methods for safety evaluation of new chemicals it is impossible to anticipate all of the conditions and permutations of use that could result in undesirable effects.

The committee noted that it is scientifically impossible to prove that a chemical is without hazard. In its conclusion the committee said:

> The level of human exposure depends on rate of application of the herbicide, balanced against the rate at which it is removed from the environment. Current patterns of usage of 2, 4, 5-T and its known fate in various compartments of the environment, including the plant and animal foods of man, are such that any accumulation that might constitute a hazard to any aspect of human health is highly unlikely.
> On the basis of these observations, it is concluded that, as presently produced and as applied according to regulations in force prior to April 1970, 2, 4, 5-T represents no hazard to human reproduction.

Despite the clear recommendation that the benefits from use of 2, 4, 5-T outweighed the possible risks, Ruckelshaus banned 2, 4, 5-T.

The scientists' majority (nine of the ten) was infuriated at Ruckelshaus' blatent disregard of their efforts. Their annoyance showed clearly in a letter to Science (November 5, 1971) triggered by remarks on the controversy made by Nicholas Wade in a previous issue:

> We, the council of the Society of Toxicology, should like to register our displeasure with the skillful selection of facts and opinions employed by Nicholas Wade to support his point of view on 2, 4, 5-T. We acknowledge the editorial policy of Science to publish minority points of view, but we cannot allow to go unchallenged such unfounded attacks on the integrity of toxicologists and other scientists in government, universities, and industries.

The 2, 4, 5-T controversy involves a fundamental issue in safety evaluation. The issue is simply whether or not demonstration of a teratogenic effect in some species of animal, at a dosage level far in excess of any possible human exposure, constitutes scientific grounds for banning the chemical. A small minority of toxicologists believe in the affirmative, but the overwhelming majority do not...

The tragedy of the whole controversy is that the EPA, by Wade's own admission, set aside the recommendations of the scientific advisory committee "in response to external pressure" and continued the ban on 2, 4, 5-T. In so doing, the administrator of EPA has done a disservice to his own scientific staff and has shaken the confidence of toxicologists in the scientific integrity of his agency. Must the majority membership of a scientific profession resort to "external pressure" to obtain sound judgment in government, rather than being permitted to devote full time to the investigation of real health hazards?

The letter was signed by nine members of the Society of Toxicology.

It is obvious that the decision to ban 2, 4, 5-T was made against the consensus of informed scientific opinion and primarily on political grounds. The bureaucratic response to rocking the boat is to beach the boat and go nowhere. It is the simple, more conservative solution, but often disastrous in the long run.

12

THE ALASKAN PIPELINE

Environmentalists and politicians have raised a tremendous brouhaha over the proposed construction of a pipeline for bringing oil from the North Slope of Alaska to the ice-free port of Valdez on the southern coast of Alaska (see map). Nature lovers have argued in some of their most passionate statements that the pipeline would cause irreparable damage to a natural wonderland. A Sierra Club editorial states that "...because of the new-found petroleum fields of the North Slope, economic values threaten to ride roughshod over all other values, human, social, cultural, wilderness, wildlife, and scenic."

In "Why the Trans-Alaska Pipeline Should Be Stopped," Congressman Les Aspin wrote, "Finally, we must all realize that the Alaskan wilderness and the way of life of many of the Alaskan natives may be destroyed if the wrong decision is made on whether to construct the trans-Alaska pipeline, and that this decision is an important one not only for Alaskans and the oil companies but for all Americans."

What are the real dangers to humans and wilderness from the proposed pipeline? A basic question is whether or not it is reasonable to expend great efforts and resources to forever keep deserted areas unpopulated and in their pristine state. What is the value to man, not just the occasional naturalist, of the desert flower that blooms unnoticed? What is the value to mankind of a ton of tundra ice? There are, of course, exceptions to most rules and to me it seems obviously desirable to leave as much open space and parkland as possible near populated areas. But this is hardly the case in Alaska, with twice the area of Texas. Alaska has 586,412 square miles and 300,000 inhabitants, about one person for every two square miles. Compare this with New Jersey, which has over 7,000,000 inhabitants living at a density of over 900 persons per square mile. Of New Jersey's 7,836 square miles, more than 60 percent is lake, forest, and farm, it is difficult, therefore, to conceive of New Jersey as overcrowded. What, then, can one say about Alaska's use, or disuse, of land? To 99 out of 100 Americans, Alaska might just as well be on Mars for all the good it presently brings them.

THE OIL

In mid-1968, after years of exploration, a large oil field was found in Prudhoe Bay, on the northern coast of Alaska. The field was discovered by geologists from the Atlantic Richfield Company and the Humble Oil and Refining Company. Considerable acreage adjacent to the Prudhoe Bay field is owned by British Petroleum. In 1969, the state of Alaska leased North Slope acreage to seven oil companies for $900 million.

Prudhoe Bay is situated on the Arctic Ocean, some 390 miles north of Fairbanks and 150 miles southeast of Point Barrow. The bay is locked in ice most of the year, 800 miles north of Alaska's southern coast. This oil and gas discovery area in northern Alaska may be one of the largest petroleum accumulations in the world and has resulted in the coining of a new English word, Prudhoemania, to describe the ditherings of some oil executives as they anticipate profits.

TANKERS OR PIPELINES

Three methods have been proposed for transporting Alaskan oil to markets in the United States. One system would carry the oil in specially designed icebreaker tankers through the Arctic-Northwest Passage to eastern markets. A second proposed system would use as yet unbuilt and untried nuclear powered submarine tankers that would run under the Arctic ice. This alternative is the most costly and speculative of the three proposals. The third system would pump the oil through a pipeline from Prudhoe Bay to Valdez on the south coast. From there it would be carried to markets by tanker.

The first proposal was tried by Humble Oil in a $40 million experiment that ran the S. S. Manhattan, an icebreaking tanker, through the Northwest Passage. In September 1969, on a pioneer voyage to test the route's feasibility, the 1,005-foot specially strengthened vessel, (the largest tanker ever built in this country), sailed from Philadelphia. She plowed her way through the ice north of Canada in a 4,500-mile trip, accompanied by two icebreakers and numerous helicopters.

The experiment proved an economic failure. Though the Manhattan got through the ice floes, she was damaged by icebergs. Under the best of conditions, the Northwest Passage would probably be unusable most of the year. In any case, Humble announced that it was abandoning its plans for icebreaker tankers and would rely on pipeline delivery of oil to Valdez.

Figure 12-1

Figure 12-1   Northwest Passage of the S.S. Manhattan

THE TRANS-ALASKA PIPELINE PROJECT (TAPS)

The Alyeska Pipeline Service Company proposes to transport Alaskan oil to the U.S. West Coast through a 48-inch pipeline from the Arctic Slope south across 800 miles of wilderness to the port of Valdez at a cost of $2 to $3 billion. The initial estimate was $900 million. It will be the largest private capital investment for a single facility in history. Tankers would carry the oil from Valdez to West Coast and other ports. As presently planned by a consortium of oil companies, the Alaska Pipeline will require a construction time of a minimum of three years for its initial stage. It will employ an estimated construction force of 5,000 to 10,000 men. The average pipe joint lengths are 50 feet. The pipe's wall thickness is about half an inch. The system's initial capacity is expected to be 600,000 barrels per day (bpd). The Alyeska Company has been studying the TAPS line for three and a half years and TAPS will be the most highly engineered pipeline ever proposed, partially as a result of environmentalist opposition.

Though the problems of building a pipeline through the permafrost region are great, they are not insurmountable. Three proposals have been suggested; each has some drawbacks (Deason, 1970). In one system the pipe would be buried in gravel berm two to five feet above the permafrost. To insulate the permafrost for 100 miles with gravel would require "grinding up a couple of peaks" of the Brooks Range. An alternate proposal is to put the pipe on wooden piling. This would require an enormous amount of lumber. For example, one pile per five feet of pipe would require over 100,000 piles in 100 miles. Pipelines supported on piles present a peculiar hazard to caribou since these animals have never learned to jump over obstacles. A third unlikely construction proposal, suggested by the Belmos Corporation, would string the pipe in the air between upright posts like a telephone wire.

The TAPS route starts near Prudhoe Bay and crosses the North Slope, roughly paralleling the Sagavanirktok River. Then it runs over the Brooks Range by way of 4,700-foot-high Chandalar Pass. The line would then cross the Yukon River near Livengood and go southeasterly from the Yukon, passing east of Fairbanks. It would next turn southward, paralleling the Richardson Highway from the Alaska Range and through the Copper River basin, across the Chugach Mountains, and through Keystone Canyon to the southern pipeline terminal at Valdez.

Construction of the line will be difficult and costly because temperatures range from 90° F. to minus 70° F. Blizzards and severe weather conditions often idle men and machinery for weeks. An estimated 80 billion cubic yards of sand, rock, and gravel would have to be used in ditches, roads, airstrips, and foundations for tanks and buildings. In areas having high moisture, corrosion is a problem. Difficult logistic problems face the engineers in storing, hauling and stringing pipe and more than 500,000 tons of line pipe, hundreds of valves and fittings, tanks, pumping units, and other equipment. Much of this equipment must be transported along an 800-mile route that has no roads for half of its length.

In general, Alyeska plans to bury the pipeline in areas where the soil can be heated without pipe settlement or damage to the ecology. About 100 miles of the Alyeska route is over permafrost; the Alyeska pipeline group sought a route in which as much uninsulated pipe as possible could be buried without permafrost degradation. In Canada, the U.S.S.R., and Alaska, engineers have built roads, airstrips, railbeds, and buildings on permafrost for years with no adverse effects.

A number of other pipeline routes through Canada have been proposed for bringing oil to midwestern markets (see figure 12-2). The total cost of the Canadian project should be about $18 billion. Incidentally, the Canadian route would traverse much more permafrost than would the Trans-Alaska pipeline. A Trans-Canada pipeline would be about 1,600 to 3,200 miles long and would cross 12 major rivers.

The Sierra Club and other nature groups have brought legal action to stop pipeline construction. They claim that a heated pipeline will thaw the permafrost and the resulting crevices would break the line and thus pollute the environment.

Figure 12-2   Proposed Alaskan (solid line) and Canadian (hatched lines)
Oil Pipeline Routes from the North Slope. The
Northernmost Range (stippled) is continual Permafrost.

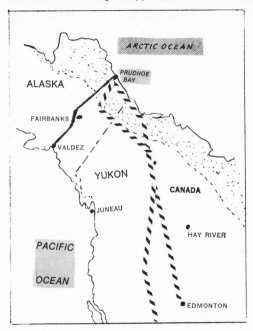

Another argument against building the pipeline revolves around possible earthquake damage to the pipeline, since southern Alaska is prone to quakes. The Alyeska line is designed to shut down automatically in case of a rupture. Welded steel pipeline has an inherent flexibility that allows it to bend rather than rupture when subjected to earth movements. The TAPS line is designed to withstand an earthquake of an 8.5 Richter Scale intensity. No electrically welded oil pipelines were broken by the severe 1971 Los Angeles earthquake.

PIPELINES COMPARED WITH OTHER TRANSPORTATION METHODS

Comparing pipeline transportation with other methods of moving oil, one finds that pipelines cause much less disturbance to the environment than almost any of them. Roads, railroads, and tankers all present greater potential hazards to the environment than a pipeline that would be buried and unobtrusive for most of its route.

210

Oil pipelines are the safest form of bulk transportation in the United States. Considered on a ton-mile basis, pipelines are 1,400 times safer and less polluting than trucks and 500 times safer than railroads. They are also more economical, carrying oil at one-fifth the price charged by the railroads, and they require less expenditure of energy than any other method.

Unlike wheeled transportation, pipelines do not throw pollutants into the air or water. Possible damage by oil to terrain is much more controllable than would be an equivalent spill on water; a fully loaded mile of 48-inch pipeline contains about 2,000 tons of oil compared to 200,000 tons of oil carried by a modern ocean tanker. And pipelines can be constructed to shut off the flow when a leak occurs, so that relatively little oil would spill on the ground.

THE WORST POSSIBLE DISASTER

Imagine for a moment the worst possible disaster that could befall a trans-Alaskan pipeline. Suppose there is an earthquake and the line breaks in several places. What would happen? Some oil would flow onto the ground in the vicinity of the breaks. With proper shut-off mechanisms, the spilled oil could be limited to a few tons per break. If the air temperature is below freezing, which is the case most of the time, the oil would not flow far, because it would congeal into a wax-like mound no higher than the level of the pipeline. The mound would cover a relatively small area. With the first snowfall, the mound would be covered and remain dormant. It is difficult to conceive of any great damage to the environment from this "worst possible disaster." When the probability of small ecological harm to a deserted area is weighed against the economic benefits of a ready source of oil, a reasonable man must conclude that the pipeline should be built.

In January 1971, after exhaustive study, the Interior Department released an environmental impact statement recommending that a right-of-way across federal lands in Alaska be granted for construction of the Alyeska pipeline to carry North Slope oil to the seaport of Valdez.

Some environmental groups still strongly opposed the project. Public hearings were held on the pipeline question by the Interior Department on February 17 and 18 in Washington, D.C., and in Anchorage, Alaska, on February 24 and 25, 1971. Alaska's governor said that the state would face bankruptcy if the pipeline were not built and it would be a great help to the native Eskimos and Indians if it were built. This later claim was disputed by the spokesman for the natives. At the time of the hearings Canadian officials were actively seeking to persuade the oil companies to route a pipeline through their country, since Canada suffers less from environmental hysteria than does the more affluent United States. Construction is still stalled.

How much wilderness will the pipeline and all its auxiliary equipment occupy? How much cold wasteland is involved? Less than 15 square miles out of Alaska's 586,412 square miles.

The main environmental argument against the Alyeska pipeline is that it may damage the tundra by exposing the permafrost. Permafrost covers an area of 130,000 square miles, about 20 percent of Alaska's northernmost regions. The pipeline will pass over at most 200 miles of permafrost in a path 50 feet wide, taking up at most two square miles of the 130,000 square miles. But life will grow on the ground around and beneath the pipe so that considerably less than two square miles of the 130,000 available will be disturbed.

AND ON INTO THE COLD NIGHT...

In January, 1972 Secretary Rogers Morton of the U.S. Department of the Interior was prepared to issue a permit that would allow the Alyeska Pipeline Service Company to start building the TAPS pipeline. The Department had prepared what it hoped was the final environmental impact statement and was ready to present it to the President's Council on Environmental Quality. But three anti-pipeline groups had filed arguments that claimed that Interior had not fully complied with the National Environmental Protection Act, requiring that it discuss alternative sources of energy in relation to the North Slope oil field. A United

States Federal Appeals Court upheld the anti-pipeliners and the Department went through the time-consuming process of listing five alternative energy sources and showing why Alaska oil was more desirable than any of them.

Then another ruling came down from the court requiring that Secretary Morton's department had to pass the revised statement around to all other agencies interested in the Alaskan matter and wait 30 days for their comments.

On August 15, 1972, Federal Judge George L. Hart, Jr. dissolved the injunction that had blocked construction of the pipeline since April 1970. Judge Hart rejected all the thousands of pages of legal arguments put forth by a consortium of environmental groups as one of the first tests of the National Environmental Policy Act. But this decision will not start construction of the pipeline: the environmentalists are appealing the decision and the case will probably wind up in the Supreme Court. In any case, the Alyeska Pipeline Service Company has stated that it would not begin construction before the legal arguments were settled in the courts once and for all.

The courts again blocked construction of the TAPS line in February of 1973, causing The New York Times to gloat in an editorial which compared the proposed pipeline project to the South Sea Bubble Swindle. A letter on the subject from H. H. Holman appeared in The Times on March 4. Mr. Holman is an Oklahoman petroleum engineer with no financial interest in the North Slope.

> ...Petroleum scientists are infinitely more reliable than the environmentalists whose ad in The Times opposing the pipeline confused barrels and gallons, introducing a 'trifling' error of 4,200 per cent...
>
> Can you seriously believe that a single pipeline, built with our best technology and sparing no expense, can significantly damage an area twice as big as Texas? Oklahoma and Texas are crisscrossed by literally hundreds of pipelines of all kinds...and one seldom hears of any of them...

213

The proposed pipeline is the most practical way of utilizing these reserves, and can be built and operated without significant ecological harm.

The increase of $2 billion in the price of the proposed trans-Alaska pipeline indicates quite clearly that one result of environmentalist attacks on raw-material extractors will be higher fuel prices, and possibly fuel shortages. Perhaps a severe power shortage, a lack of gasoline for America's one hundred million polluting automobiles and a curtailment of electricity for the air conditioners and washing machines of environmentalists might have a salubrious effect on the Alaska pipeline debate.

13

ELECTRICAL ENERGY AND THERMAL POLLUTION

Not only the lights dimmed in Great Britain towards the end of the coal miner's strike. During the winter of 1971-72, British cities were forced to dump millions of gallons of untreated sewage into rivers and estuaries; water treatment as well as sewage plants were knocked out by lack of electricity; electrified trains couldn't run; and millions of workers were laid off as hundreds of industrial plants were forced to close down.

In the 20th Century, electricity has replaced millions of individual sources of man-made pollution: kerosene lamps and open fires by electric lights; combustion engines by electric motors; coal and oil stoves by electric ranges; dirty furnaces by electric heaters. Electricity provides power for sewage and garbage treatment, for cleansing dirty air and water, and for recycling wastes. It drives electric trucks, buses and trains without directly generating air pollution. New anti-pollution devices now being developed depend on it.

Low-cost power is needed to improve and protect the environment. In the words of Lee A. DuBridge, presidential science advisor, in testimony before the Joint Committee on Atomic Energy, "An abundant supply of low-cost energy is the key ingredient in continuing to improve the quality of our total environment." It is ironic that only wealthy industrial nations that can generate large quantities of cheap power may be able to maintain clean environments in the future. Conferees at the 1971 Eighth World Energy Conference in Bucharest, Romania, concluded that more, not less, power is needed to reduce pollution. They directly contradicted environmental extremists who claim that "all power pollutes"; extremists who would like to go back to machineless way of life.

215

The fight against pollution requires power, lots of it, to recycle wastes, provide pollutionless mass transportation, and energize pollution control devices such as precipitrons, water pollution control plants, and air scrubbers. Growing food economically for the world's expanding population requires more, not less energy. Pumping irrigation water, providing illumination and setting up aquaculture systems require energy. Where will the energy come from?

Nuclear reactors now provide less than 2 percent of the United States' total electrical power. But by 1980 "nukes" will be producing 25 percent of our energy, and by the year 2000 over 50 percent of all electrical energy will come from nuclear sources. Instant environmentalists, often with the help of an uncritical press, have tried to create the impression that nuclear power plants are much more harmful to man and his environment than fossil fuel plants. But in general, reactors will be located away from population centers, probably on islands built in the ocean, and electricity will travel to the cities through high-voltage submerged power lines. Instead of the boiling water reactors being used in the 25 or so atomic generators now in use, the plants designed after 1980 will probably be of the fastbreeder or fusion types, both of which are more efficient and environmentally desirable than present-day reactors.

Three major environmental problems arise from the generation of electric power: the misuse or uglification of the landscape; possible damage to water resources by the discharge of thermal effluents; and the danger of radiation from nuclear power plants. In this section, we consider one of the major ecological charges against power plants in general, and nuclear plants in particular, that they generate inordinate amounts of waste heat that is ruining, or will ruin, the environment. In the following chapter, we consider the environmental issues of radiation from nuclear power plants.

THERMAL EFFLUENTS: COOLING IT
The use of electricity in the United States is projected to double in 10 years, as it has during the past 10 years.

216

The number of generating plants will increase. Each power plant has boilers that must be cooled. Ambient water or air is used to cool the boilers, and it is returned, still hot, to the environment. The introduction of waste heat into the waterways, oceans and atmosphere is called thermal pollution or "thermal enrichment" by the public relations department of Consolidated Edison.

Steam power plants now require about 80 percent of all cooling water used by industry. Hydroelectric plants, which now generate 20 percent of all power, do not create warmed effluents, but the percentage of hydroelectric plants will decrease in the next 20 years as fossil fuel and nuclear power plants proliferate. By the year 2000, all of the fresh water in the United States will be required for cooling power plants if present trends continue.

Nuclear light-water plants now being built waste about 70 percent of the heat energy generated by their atomic fissioning reactions. The heat usually is carried away by flowing water. Fossil fuel plants are about 40 percent efficient, so that for every two calories of heat released from oil, coal, or natural gas, less than one calorie is turned into usable electricity, and more than one is fed into the environment. Of the waste heat from a fossil fuel plant, $10°$ percent to $20°$ percent is discharged up the smokestack into the air.

Water has a high heat capacity: it can hold a relatively large quantity of heat for a given mass. Water is a cheap coolant when available. In power plants, ambient water is pumped through the condensers and is rejected some 10 to 30 F warmer than when it entered.

Because warm water is lighter, less dense than cold water, it rises to the surface in a plume and floats. Stream currents disperse the plume, spread it out, and mix it with cooler waters (see figure 13-1). Air temperature and wind speed also affect the cooling and mixing process in the receiving water.

Waste heat can be dissipated in a number of ways:

1) Water can be run one time from a water body through the condensors, the so-called "once through" method. This is usually the least expensive technique and the one favored by most power engineers.

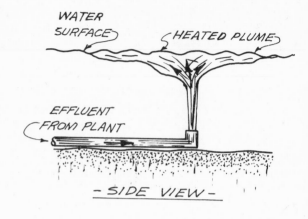

WATER SURFACE

HEATED PLUME

EFFLUENT FROM PLANT

- SIDE VIEW -

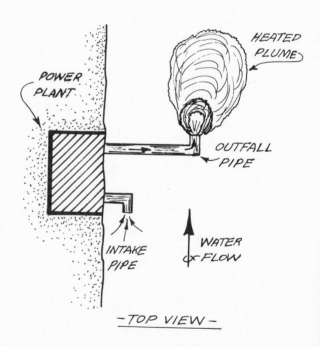

POWER PLANT

HEATED PLUME

OUTFALL PIPE

INTAKE PIPE

WATER FLOW

-TOP VIEW -

Figure 13-1   Plumes of Heated Water from Power Plants rise to the Surface (top) and Spread in the Direction of the Current (bottom).

218

2) The heated water can be run into ponds or lagoons, where evaporation and conduction cool the water. Such pools require enormous surface expanses to operate properly; for instance a 2,000 megowatt nuclear power plant requires a 2,000-to-3,000-acre cooling pond, depending on atmospheric conditions: a cool, dry climate necessitates a smaller pond than a warm, moist one.

3) The heated effluent water can be run through cooling towers, large tube-like structures in which the water is cooled by contact with the cooler ambient air or by evaporation. Towers are much in vogue by some environmentalists as a method of eliminating "once through" techniques that on occasion have killed fish.

Cooling towers are relatively expensive and can add as much as 30 percent to the cost of nuclear power plants. They are ugly or not depending on one's taste for phallic monuments, but they are certainly not "natural" looking. Water vapor from a cooling tower may cause fogs, or icy roads and icy power lines. Large volumes of water vapor added to the atmosphere could change the climate near these towers.

A rain of salt may be a problem if the water used for cooling comes from the ocean or from a salt water estuary. The AEC may order the Baltimore Gas and Electric Company to build two 530-foot high cooling towers for the Calvert Cliffs plant instead of using water drawn from Chesapeake Bay. A 1972 report from the utility company to the AEC estimated that the proposed cooling towers could deposit "as much as 102 tons of salt per square mile per month on adjoining lands at a distance of about one mile." Incidentally, the tower would add $108 million to the cost of the plant compared with once-through cooling.

Loss of waters from cooling towers could cause a problem for towers on inland lakes and rivers; a 2,000 megawatt plant would loose 14,000 gallons of water per minute through a cooling tower. Obviously such losses could affect stream flow, as well as increase the humidity in the vicinity of the tower. The Indian Point complex in New York, when drawing full power, would require almost all of the Hudson River's slack water flow.

219

AGAIN THE OCEAN

In 1971 the New York State Society of Professional Engineers, in a report on the energy problem, recommended that nuclear plants on the East Coast be sited only in or on the Atlantic Ocean. Power utilities are already finding that they must build on the ocean in order to meet the thermal effluent standards set by the federal government. Discharge of heated effluent into a lake or estuary will raise the temperature of the receiving waters in a way directly related to the relative quantities and temperatures of the waters involved: discharging 100 million gallons of heated water at $98°$ F. into a lake containing 100 million gallons of water at $90°$ F. will result in a mixed 200 million gallons of water at $94°$ F. The smaller the receiving volume, the greater will be the resulting rise in temperature.

But the ocean, if properly utilized, presents an enormous mass of moving water and can act as an almost infinite sink for man's waste heat. California's shore currents are strong, and the shelf slopes rapidly so that waste heat piped for years into the Pacific has dissipated rapidly and the ocean water temperature has changed little. Studies have failed to reveal any detrimental effects resulting from offshore disposal of thermal effluents. California power plants expect to discharge effluent waters $20°$ F. higher than ambient despite Environmental Protection Agency guidelines, which would appear to prohibit such a temperature differential. In any case the receiving ocean waters will not rise to the $90°$ F. level deemed critical for some marine species.

Florida State officials, faced by environmentalist outcries against "thermal pollution," proposed guidelines for power plants in December 1971 along the lines suggested by the EPA. The guideline proposed by the Florida State Pollution Control Board would limit the rise in temperature of the receiving body of water to only $1.5°$ F. in summer. These new rules are likely to force some power companies to scrap existing plants and plans and build new generating plants with access to the Atlantic Ocean with its vast quantities of cool, moving water. The alternative would be cooling towers or cooling ponds, neither of which is inexpensive or necessarily environmentally desirable.

In April 1973 the Marine Technology Society sponsored a meeting in Florida to consider another way to cool power plants. The Conference on Offshore Nuclear Power Plants heard papers on the proposal by the Westinghouse, Tenneco and Jersey Utility corporations to build floating nuclear power plants in the Atlantic Ocean three miles from the Jersey littoral. Other groups, including Con Ed and the Electric Whale Company, are investigating offshore power plants.

While thermal effluents from power plants will increase the temperature of water bodies that are relatively small in relation to the volume of water expended by the power plants, their effect on larger lakes is likely to be small. Lake Michigan, for instance, is a big body of water. Argonne National Laboratory scientists studied the lake and the projected growth of power plants on its shores and concluded that "lakewide effects of man-made thermal discharges into Lake Michigan are negligible and will continue to be for the rest of this century." Calculations by Asbury and others indicate that the temperature of the Lake will probably rise one degree fahrenheit by the year 2000-a rise that should not cause much change in the flora and fauna of the lake when the effluent is properly dispersed.

POTENTIAL DELETERIOUS EFFECTS OF HEATED EFFLUENTS

As the temperature of water rises, its capacity to hold dissolved gases such as oxygen decreases and in the summertime, surface waters sometimes become oxygen-poor naturally and cannot support fish and other animals. Thermal effluents accelerate the depletion of oxygen from water and thus may be responsible for fish kills. But in a large water body, marine animals will usually leave an area when conditions for survival are poor and will migrate to another part of the lake or ocean. In cold weather, many marine animals survive because of the heated effluents from power plants; an "unnatural" but beneficial condition.

The capacity of a water body to assimilate wastes depends on the amount of oxygen in the water: a heated stream can assimilate less sewage waste than a cooler stream of the same size. Often wastes will work in

221

conjunction with warmed water to deplete it of oxygen. In 1968 and 1969, Dr. George Claus and I conducted experiments to see if bivalves would survive in the East River, which is really an estuary along the east coast of Manhattan Island. We suspended baskets of oysters, clams, and mussels into the river. The experiments started in the fall, continued for about a year, and were interrupted only by humans who, out of curiosity I presume, emptied our bivalve baskets on the South Street Seaport dock. We found that the shellfish did well most of the year in East River waters. Only during the hot summer months did we note a partial die-off of some of the larger oysters, probably due to temporary oxygen depletion.

Many aquatic organisms are sensitive to temperature changes; a temperature increase or decrease will disrupt their life cycles, sometimes to the point where natural reproduction is adversely affected. The female oyster , for instance, hibernate when the water cools in the winter. As the estuarine waters warm in the spring, she undergoes physiological changes. With the right temperature, which varies for different species and different locales, the female commences to disperse millions of eggs into the water and the male disgorges hundreds of millions of spermatoza. Some of the spermatoza find and fertilize some of the eggs. The fertilized eggs become oyster larvae, which swim about for a few days, then settle down on a hard substrate to grow, mature, and repeat the life cycle. If the water is prematurely warmed by thermal effluents or by some natural disruption, the female oyster's prodigious efforts are likely to be in vain, since her millions of young larvae will not find sufficient algae in the water on which to dine. Aside from disrupting an animal's life cycle, high temperatures can damage or kill aquatic fauna directly if the animals are not able to move away.

The marine environment is an extremely complex one in which many aquatic biological and physical processes occur, and their relative importance to the survival of marine organisms are not completely understood. This includes temperature changes. Some biologists have theorized that warmer water might cause an overall increase in plankton production, which should then lead to an overall

222

improvement in commercial and sport fisheries for that area. But laboratory experiments show that summer plankton live close to their upper limit of heat tolerance and could not survive at temperatures even a few degrees higher.

Marine organisms often become adapted to the warmed water around power plant effluent outfall pipes. A number of cases have been reported where fish died in the winter because the plant was shut down for one reason or another. In these cases a decrease in temperature caused the fish kills. Dams also cause a decrease in water temperature by forming lakes with large surface areas. Evaporation from the surface waters cools the entire water body and brings about a different ecological system than existed before the building of the dams.

Thermal pollution is currently destroying the fishing grounds of trout, salmon, whitefish and freshwater herring by reducing the oxygen content of the water. Only a small temperature change in a stream may make it lethal to sensitive trout: though the slight increase may occur once a year for a mere two hours on a single day, the water-course can lose all its trout. Three parts per million of dissolved oxygen is usually considered the lower limit for many fish; if the oxygen drops below that level the fish will die or move away. Higher levels are preferable to sustain normal fish activity. Fish are coldblooded animals and their body temperature usually differs only slightly from that of the ambient water. Temperature increases can cause fish kills in a number of ways: by accelerating enzyme reactions so that normal enzyme activity is inhibited; by melting body fats; by coagulating cell proteins; or by reducing the permeability of cell membranes.

Waste heat may also adversely affect the growth of algae and other water plants whose development and life cycles are temperature-dependent. Temperature changes may eliminate beneficial species and result in the establishment of nuisance species. However, the reverse may also be true. Since algae are at the base of the marine food chain, any change in quantity or species is likely to affect the quantity and nature of animals higher up on the food

223

web. In certain limited cases, which also occur naturally, thermal changes can prove hazardous to humans. At a meeting at Portland, Oregon (Krentel, 1968), Dr. Albert H. Stevenson, assistant surgeon general and chief engineer of the U.S. Health Service, warned that warm water discharges might increase populations of certain poisonous planktonic organisms that are consumed by shellfish. Measurements in Puget Sound show that during the summer a rise in temperature as small as one degree fahrenheit could trigger a bloom of such organisms and in consequence, an epidemic of shellfish poisoning. However, poisonous plankton such as red tide dizoflagellates bloom naturally in many ocean areas when the water is warmed by the sun. Man does not yet have the ability to control these natural blooms.

BENEFICIAL USES OF WASTE HEAT

After describing some of the horrible things that heated effluents may do to the environment, it is only reasonable to consider what good may be done with the waste heat rejected by power plants. A great deal of effort has gone into investigation of waste heat utilization for economic as well as environmental reasons. Electric power companies are now wasting over 50 percent of the heat energy that they generate by burning fossil fuels or fissioning atoms; this amounts, literally, to flushing away billions of dollars annually. Up to now it has been cheaper to throw the heat away than to try to use it. But some uses have been developed, and more are sure to follow.

One obvious use of waste heat from power plants is to heat buildings, and there is no technical reason why heated water or steam cannot be piped to buildings in need of warmth. Steam is used extensively to heat buildings in downtown Manhattan. Unfortunately, in the summer, air conditioners cause a peak rise in electricity demand, thus increasing the amount of waste heat at the power plant when it's least needed. Using heat to warm or cool cities is not now an economic reality in most existing towns, but if the electric power monopolies were forced to use more of their profits in research along these lines I suspect that this and other uses of waste heat would be usefully employed rather quickly.

Vitro Corporation of America has been experimenting with the use of hot water from a Weyerhauser Company pulp and paper plant in an attempt to increase agricultural yields in Oregon's Willamette Valley. Hot water from the plant is piped to orchards and crop lands and distributed by a grid of smaller pipes and spray nozzles. In the summer, plants are cooled and watered by the spray, which is raised so that it is cooled by the air before it hits the plants. During the cooler seasons the hot water protects the trees and plants from frost. Plants grow faster due to the better regulated soil temperature.

A number of programs have been developed to use waste heat for increasing the production of marine organisms. This is a logical way to try to use heated effluents, since they are usually liquid to begin with. Fish thrive best in waters with optimum temperatures specific to each species. The American oyster for instance hibernates when the temperature drops to the near-freezing levels that occur in waters during the winter along most of the Atlantic Coast. Therefore it takes a Canadian oyster six years to reach edible size, a New England oyster four to five years, a Virginia oyster three to four years. But if the temperature can be optimally controlled throughout the year, and the oysters properly fed, they can mature in less than a year. For several years the Long Island Lighting Company (LILCO) has been pumping warm water from its Northport Power Plant into a lagoon in which young oysters were kept in an attempt to extend the growing season and convince the vocal environmentalists on Long Island that LILCO loves the environment, the results having been encouraging. Japanese technologists have been able to grow shrimp throughout the year instead of only from May to October by using heated effluents from power plants. The economic incentive is high since live shrimp sell for up to $15 per pound in Japan. The British have been growing plaice, a flat fish, in artificially warmed Scottish waters, and have found that these fish mature much faster in heated lagoons than in unheated waters.

In some areas in Great Britain, near thermal outfall pipes, American hard clams are thriving where they did not previously exist.

Fresh water catfish prefer warm water. With the help of heated effluents, the catfish farmer can grow fish throughout the year, rather than only seven or eight months. At least three power plants in the mid-South (where catfish are widely eaten) are engaged in supplying heated effluent for catfish farming.

In addition to the above uses, heated effluents can be used to desalt sea water. Waste heat can be used to treat sewage and sewage sludge by speeding up biological decomposition and sterilizing the effluent. Waste heat can be used to keep lake and river routes ice-free in the winter, to heat greenhouses, and to defog airports and deice runways. Potential beneficial uses of thermal effluents abound: what is lacking at the moment is the will and the money to put these ideas into practice. America is still too fat and wealthy to worry about using all of its wasted energy. But it is only a matter of time before many of the projects mentioned above are built. Waste heat has too much value to let it dissipate aimlessly into the environment. Today's thermal "pollution" is a great untapped future resource.

14

DEATH FROM RADIATION AND NUCLEAR ENERGY

Hiroshima's mushroom cloud introduced the Atomic Age to the consciousness of mankind-Hiroshima with its malformed babies, its deformed maidens, its thousands of innocents killed in a single blast of radiation and heat. It is small wonder that mankind in the 1970's harbors deep-seated fears regarding nuclear power plants. The public has been inundated by books such as Population Control Through Nuclear Pollution, The Careless Atom, and Perils of the Peaceful Atom, and by strong statements by a relatively few scientists and their followers on death from radiation. Consequently the public has developed fears that have been expressed in public action that have held up the construction of many power plants. The Calvert Cliffs decision of 1971 should strengthen the position of those who oppose the construction of new plants on environmental grounds. This landmark decision forced the Atomic Energy Commission (AEC) to reject a permit for a large nuclear power plant on the shores of Chesapeake Bay because of an inadequate environmental impact statement.

Today's nuclear power plants use uranium pellets as fuel. The pellets are usually enclosed in thin-walled, vertical fuel rods of which there are likely to be over 20,000 in a typical 500-megawatt (MW) plant. The rods are 1/8 to 1/2 inch in diameter. As the uranium fuel in the rods fissions (splits), a small quantity of atomic mass is converted into heat energy, radiation, and new elements. The fission products include radioactive strontium 90, cesium 137, iron, krypton and iodine. Water is pumped through the closely stacked rods to remove and transfer the heat to where it can be used to produce steam to turn a turbine that generates electricity. A small amount of radiation can, and has, leaked out of the 25 or so nuclear power plants now operating in the United States via cooling waters that have interacted with faulty fuel rods; via smoke stacks as radioactive gasses; and during the disposal of spent, but still radioactive, fuels. But this radiation, as we shall see, is relatively insignificant.

All reactors now in use utilize water to carry heat from the fuel rods. The boiling water reactor (BWR), also known as a light water reactor (LWR), allows water to boil within the reactor chamber and thus produces steam at the top of the chamber, steam that is then piped to the turbines. Pressurized water reactors (PWRs) maintain the water contacting the fuel rods at a high pressure that prevents it from vaporizing. The heated, high-pressure water runs through a heat exchanger where it heats water in second heat exchange system. The water in the secondary system vaporizes to steam which turns the turbine blades. In the PWR system the radioactive section of the power plant is isolated from the other sections more effectively than in the BWR system.

By 1990 more than 50 percent of U.S. electric power will probably be generated in nuclear plants, according to nuclear enthusiasts. Future nuclear reactors probably will be of the fast breeder type, or depend on fusion rather than fissioning of atoms for release of energy. The fast breeder reactor (FBR) requires less fuel than the BWR or PWR. The breeder is more efficient and thus needs less cooling water. It is less likely to heat up the environment because it produces less thermal effluent. It also produces less radiation. Unlike fusion reactors, FBR's are based upon relatively well-established technology. An enormous amount of uranium fuel can be saved by use of FBR's rather than light water reactors, thus conserving natural resources.

The AEC has strenuously promoted breeder reactors, particularly the liquid metal fast breeder reactor (LMFBR). President Nixon has asked Congress for money to push LMFBR development, which is expected to cost up to $4 billion. However, there is considerable doubt that the AEC's glowing estimates of cost savings and safety are accurate. T. B. Cochran, a physicist with Resources for the Future, an independent Washington study group, analyzed the AEC cost-benefit analysis for the LMFBR and concluded that it depends on four dubious assumptions: 1) that demand for electric power will grow by the year 2000 to 10 trillion kilowatt hours rather than half of that, as now seems more likely; 2) that the interest rate on capital will be 7 percent, rather than the more than 9 percent now being charged by bankers; 3) that uranium fuel for conventional LWR reactors will be in short supply and very costly by the 1990's; and 4) that there will be no unforseen safety problems which may well increase the cost of reactors. Serious questions as to the safety of LMFBR's abound, however, and the promotion of this reactor to the neglect of other forms of power generation and energy conservation appears to me to be imprudent. Breeder reactors operate at higher temperatures than do LWR's and are much more difficult to control. A potentially disastrous side effect of LMFBR's is that they generate plutoninium, a substance that can rather easily be made into atom bombs.

RADIATION HAZARDS AND BENEFITS

Hazards associated with radiation have been known for many years because of the widespread use of x-rays, radium and other radiation-based technologies. Strict governmental controls on radiation exposure were imposed long before other environmental contaminants were even considered. Standards for protection against ionizing radiation are comprehensive. The standards suggested by the International Commission on Radiological Protection and by the Federal Radiation Council have been adopted by the AEC, the U.S. agency in charge of enforcing standards in the nuclear field. The ideal situation would allow no release of radioactive wastes to the environment, but the AEC and some components of the nuclear reactor industry have resisted attempts to reduce radiation levels to what they consider economically and environmentally unjustified levels.

Large doses of ionizing radiation have been shown to cause leukemia and other forms of cancer, reduction in fertility and deformation of offspring, cataracts, and accelerated aging. This list of disabilities must be tempered by the knowledge that there is little real understanding of the long-range effect of low level radiation-and that some radiation probably is essential to life and growth.

In two recent European experiments, animals were isolated from natural cosmic radiation that bombards all animals, including man, that live on earth. Scientists in Czechoslovakia and in France enclosed rats and other animals in lead-lined boxes and then lowered the beasts into mine shafts to prevent cosmic rays from reaching them. The scientists found that after four generations of animals had been raised in radiation-free conditions, they exhibited a variety of malformations and increased susceptibility to disease. Experiments with insects and yeast cells shielded from natural radiation yielded somewhat similar results. These and other findings suggest that radiation is necessary to normal cell development and that man could probably not live without it. From an evolutionary point of view it is certain that life would not have evolved into its present form without the aid of genetic mutations brought about by radiation. Again, as with so many other substances, contaminants or pollutants, mankind simply does not know where to draw the line between quantities that are harmful and those that are necessary or beneficial to life.

ON THE DEATH OF INFANTS DUE TO RADIATION FROM NUCLEAR ENERGY PLANTS

A religious group in the sunny fruit and nut state gained wide publicity and adherents during the late 1960's by predicting that California would slip into the Pacific Ocean in 1970. Arguments based on technical evidence that the event was unlikely did not dissuade believers. They based their beliefs and forecasts on Biblical readings. From a scientific point of view it is impossible to prove or disprove in advance a prophesized disaster. One had to wait until 1970 ended before one had proof that the state had not fallen into the Pacific.

In 1972 I attended an environmental meeting in New York at which a young medical doctor, an associate of

Ralph Nader's, stated that a host of industrial, chemical and nuclear substances were damaging the health of the nation and were causing increased death and disease in many subtle ways. He may be right. But I asked him how he explained the contradiction that life expectancy has been rising constantly during the industrialization of America in the first 70 years of the 20th Century. "Ah," he said, "wait 30 years." How can one argue with that?

Recently, I picked up a neighborhood newspaper and read the following headline: "Scientist Fears Contamination." The article started, "The incidence of infant deaths is higher in communities with nuclear reactors. This is the opinion of University of Pittsburgh radiologist Dr. Ernest J. Sternglass..." It went on to say, "According to Dr. Sternglass, whose openly expressed opinions have confirmed long-standing medical rumors: 'to proceed in the face of such evidence with an expansion of the nuclear energy program and to install both small and large reactors in the midst of our heavily populated areas would represent a reckless gamble with the health of the nation's children for generations to come.'"

For many years Dr. Sternglass has been promoting his opinion that nuclear power plants cause a demonstrable rise in infant mortality. His appearance in our local newspaper came about because Columbia University wished to start up a small research reactor in the basement of its engineering building. A group of vociferous local citizens have opposed the reactor, and indeed there is little justification for placing the reactor in a densely populated neighborhood when Columbia owns land nearby in unpopulated areas, and other reactors are readily available to Columbia students and teachers. The legal wrangling has been going on for seven years. In April 1971 the Atomic Safety and Licensing Board of the Atomic Energy Commission denied Columbia the right to start up the reactor, the first such denial by the Board. But Columbia, with the prodding of the reactor manufacturer, decided to fight for the reactor. As a result, the anti-reactor group asked Dr. Sternglass to testify at hearings held in October 1971 and February 1972.

Almost all the scientists who have studied Sternglass' data and charges say he is wrong. Dade W. Moeller, the president of the Health Physics Society of Harvard University's School of Public Health, has said that Sternglass' allegations have been repeatedly analyzed and found wanting. In a letter to Science written after Dr. Sternglass had presented his infanticide-reactor story before the Health Physics Society meeting in New York in July 1971, Dr. Moeller wrote:

> On the third such occasion since 1968, Dr. Ernest J. Sternglass, at an annual meeting of the Health Physics Society, presented a paper in which he associates an increase in infant mortality with low levels of radiation exposure. The material presented in Dr. Sternglass' paper has also been publicly presented at other occasions in various parts of the country. His allegations, made in several forms, have in each instance been analyzed by scientists, physicians and biostatisticians in the federal government, in individual states that have been involved in his reports, and by qualified scientists in other countries.
>
> Without exception, these agencies and scientists have concluded that Dr. Sternglass' arguments are not substantiated by the data he presents...

Yet the public fear of radiation-fueled in part by spokesmen such as Sternglass, and headlined by an irresponsible press-is real and has led to postponement and abandonment of nuclear projects. Little can be done in a democracy to prevent astrologers, religious zealots and rogue scientists from broadcasting their views. Only an informed and rational public can prevent precipitous or foolish governmental decisions.

NATURAL AND MAN-MADE RADIATION

We are all well-irradiated every day of our lives. Many natural substances give off alpha, beta (x), gamma rays and cosmic rays. Life has developed on earth with this radiation as a background and a goad. It is unlikely that life could have evolved to its present state without natural radiation because bombardment of genes by particles causes mutations that undoubtedly have led to better-adapted forms of life such as man.

About 55 percent of the total radiation reaching the average American each year comes from natural sources, but the Denverite on a clear day receives a much stronger dose of cosmic rays than does a New Yorker living at sea level.

Fallout from atom bomb tests in the atmosphere account for about 3 percent of man-induced radiation. This percentage is decreasing and, short of a nuclear war that would destroy most of mankind, it should continue to decline.

About 45 percent of the total radiation hitting the average American comes from x-ray machines, TV sets in the home, and nuclear power plants. By far the greatest portion of this environmental radiation (over 90 percent) comes from dental and medical use of x-rays. TV sets are the next largest source of man-made radiation; nuclear power plants are the smallest source, representing less than 2 percent of the total.

J. G. Terrill, Jr., compared the costs and benefits for reducing radiation dosage from nuclear reactors and from medical treatments. He found that by using automatic collimators on diagnostic x-ray equipment, radiation exposure of the general public could be substantially reduced at a moderate cost per unit of exposure-$7 per unit-compared with a cost of $100,000 per unit for reduction of exposure to reactor radiation for an equivalent unit. Terrill pointed out that the population of the United States is currently exposed to 430 units of radiation from nuclear plants as against 18.7 million units from diagnostic x-rays.

Many minerals, including coal, are radioactive. Curiously enough, coal-fired power plants often emit <u>more</u> radioactive material than do nuclear plants. Natural background radiation in the vicinity of rock-bound New York is about 125 radiation units per year. According to Charles Luce of Consolidated Edison, "Inside Grand Central Station, measurements have recorded 79 to 525 additional radiation units per year, all because of the granite."

But the number of nuclear plants is scheduled to mushroom in the next 20 years from the presently operating 25 to about 450 plants in 1990. These plants release small amounts of radiation to the air and water, emissions which now constitute less than four/one thousandths percent of all man-made radiation.

The increase in nuclear plants will aggravate environmental problems such as fuel reprocessing and disposal of radioactive wastes. Persons living in the vicinity of plants have more cause to worry about accidental discharge of large quantities of radioactive materials should the reactor "run away," rather than about low level radiation. Though the probability of an accident is extremely low-less than the probability of being struck by lightning-residents in an area where a plant is planned may not wish to assume the added risk. This is the case in Shoreham, New York, where Long Island Lighting Company plans to build a nuclear power plant. The residents of the area have marshalled an impressive array of environmental lawyers and persons with scientific degrees to fight the

Table 14-1

Nuclear Power Reactors in the U.S.A. Present and near Future - 1973

	Number	Kilowatts (approx.)
Operable (not nessarily operating)	25	12 Million
Being built	52	42 Million
Planned (Reactors Ordered)	53	53 Million
TOTAL	130	107 Million

power company. An obvious, though expensive, solution is to build nuclear power plants on artificial islands, or mounted on ships moored in the Atlantic Ocean. In the long run, this solution may be less expensive than building plants on land near populated areas. Since the spring of 1972, I have been studying the design and installation of fossil fueled plants 12 miles or more offshore. Offshore fossil plants have many environmental and economic advantages compared with other power plant systems, and within a few years we should see coastal communities turning to offshore power plants as the most acceptable sources of virtually pollution-free electric power.

Radioactive wastes from nuclear plants are usually in liquid form. Some 80 million gallons of "hot" (radioactive) wastes are now in the process of being solidified by a new process that reduces the overall volume and minimizes leaching of radioactive particles into ground water. Hot wastes now are usually buried or stored in caves, abandoned mines, or other carefully selected, isolated sites where they can be monitored.

For many years the AEC has planned to use abandoned salt mines in Kansas for storage of hot reactor wastes. But a danger of contamination of underground water has frightened Kansas legislators, who have risen up against this proposal to make Kansas a graveyard for the nation's nuclear trash. Despite assurances from the AEC that no harm will result, the burial program has been stymied, perhaps indefinitely. Unless alternate disposal methods are found fairly soon, the problem of waste disposal could hold up the entire liquid water reactor plant construction program.

X-RAYS FROM MEDICAL UNITS AND TELEVISION SETS

If the average American is getting an overdose of man-made radiation, it is more likely to emanate from his television set and from medical treatment than from nuclear power plants. The major source of man-made radiation to the U.S. public comes from x-ray units in hospitals and doctors' offices. Many of the operators of these high-voltage units have not had any formal training in protecting patients from excess radiation. According to Dr. Karl Z. Morgan, director of the health physics division of the Atomic Energy Commission and a leading authority in the health physics field, "X-rays are badly oversubscribed by many doctors, and the excessive radiation that results may pose a real threat to the health of individuals and (through genetic damage) to their future children."

It is a statistical fact that leukemia is much more prevalent among children whose mothers have undergone x-ray treatment during pregnancy than among comparable mothers who have not been exposed.

Two English researchers, A. Stewart and G. W. Kneale, have shown that pregnant women who are x-rayed are more likely to have children who develop cancer than are mothers who are not x-rayed. Unborn children may be a hundred times more sensitive than adults to radiation damage.

All told some 130 million Americans went under the x-ray machine in 1970 and were exposed to 500 million blasts of radiation. Most exposures were for diagnosis of fractures, since x-rays travel through soft body tissue with relative ease, but cast a shadow when intercepted by bones. The shadow-graphs of x-rays hitting a photographic plate constitute an x-ray picture. Much x-ray use is medically unjustified: one study of 1,500 x-rays taken of 435 persons with possible head injuries concluded that only one of the x-rays actually aided the diagnosis.

Aside from unneeded exposure during diagnosis, patients are often subjected to overdoses and irradiation of parts of their bodies not under surveilance. These excesses are often due to indiscriminate use of x-rays for the convenience of doctors and administrators, faulty equipment, and undertrained x-ray technicians. The situation appears to be improving somewhat: excessive doses to patients decreased from 50 percent in 1964 to about 33 percent in 1970. Under a new law that should take effect in 1973, the Public Health Service is setting new radiation standards for equipment and for technicians.

Almost every American home has one or more television sets and every television set emits x-rays. These x-rays are produced when electrons from the cathode of the picture tube strike the picture-tube screen. The higher the voltage used to accelerate the electrons, the "harder" and more penetrating will be the x-rays. Color TV sets usually require higher voltages than do black-and-white sets and are thus more potent sources of hazardous radiation. A survey conducted in 1968 showed that 6 percent of color sets selected at random emitted radiation in excess of the standards set by the Council on Radiation Protection. Even a properly shielded set emits some x-rays that conceivably could harm humans. Children watch TV in America for an average of 20 hours per week.

RADIATION IN PERSPECTIVE

Every American is exposed to natural radiation from outer space and from radioactive minerals. Some Americans live in areas where the soil and building materials are particularly hot: residents of one Colorado town found that their homes were built on radioactive fill from nearby uranium mines. Certain areas are naturally hot: springs bubbling up through radium-rich soils have been found to contain over 500,000 picocuries per liter of water compared to the maximum concentrations permitted in drinking water of three picocuries per liter. Every American who watches TV and undergoes medical x-ray treatment exposes himself to man-made radiation far in excess of any power plant he is likely to encounter.

In September 1971, leading nuclear scientists met in Geneva under the aegis of the United Nations Atoms for Peace Conference to discuss the future role of atomic energy and radiation hazards in the world. The overwhelming consensus of the scientists was that nuclear generated power was "clean, safe and necessary." By the year 2000, after an enormous worldwide increase in the number of nuclear power plants, the total amount of radioactivity produced by the entire nuclear industry would still be only one percent of the natural radioactivity coming to earth from cosmic rays and natural radioactivity from minerals. In the public mind, radiation is still usually associated with atom bombs and cancer. Only a misinformed public frightened by fancied hazards of radiation exposure can prevent the atom from taking its place beside fire as a useful servant of mankind.

15

AH WILDERNESS!

"Where man is not, Nature is barren."

William Blake

The related myths of wild forests and vanishing species are based on a primitive, mainly subconscious feeling that forest and other pristine areas are essential to the well-being of mankind, and therefore must be kept forever wild. Some wildlifers maintain that man must preserve all existing species of wildlife or else the world store of genes will be seriously diminished, and life on earth will in some mysterious way eventually collapse.

Mystery lurks in the shadows of the deep forests, a mystery that primitive man worshipped, as do some of his descendants. The ancient Greeks thought trees the abodes of the gods. The Druids made churches of their forests; the early Aryans in Europe worshipped around the tree at midwinter as do their descendants at Christmas time. It is small wonder that myths regarding forests still linger in the minds of modern man.

It is the nature of the human animal to want to see the horizon. In medieval times, Europeans had a "horror sylvanus," a fear of the woods that was very real to them. Utterly surrounded by forests, man felt he must cut down trees and create open space for himself. And some of us city dwellers have developed a "horror concretus" which drives us back into the woods. So it goes.

Throughout civilized history, men have longed for the Forest of Arden, away from the clash and troubles of civilization. How sweet to hark back to the primeval, sheltering trees.

Under the greenwood tree
Who loves to lie with me,
And turn his merry note
 Unto the sweet bird's throat,
Come hither, come hither, come hither.
        Here shall he see
        No enemy
But winter and rough weather.

Who doth ambition shun,
  And loves to live i' the sun,
Seeking the food he eats,
And pleased with what he gets,
Come hither, come hither, come hither.
        Here shall he see
        No enemy
But winter and rough weather.

Thus in <u>As You Like It</u> sang Amiens to celebrate the forest-lovers creed.

FORESTS IN AMERICA TODAY

"And out of the earth made the Lord God to grow every tree that is pleasant to the sight."

Genesis 2:19

Three hundred years ago about 42 percent of the area making up the 48 contiguous United States was covered by forests. Today this land is still approximately one-third timbered, although not all of the present day trees are virgin growth. Of the more than 500 million acres of commercial U.S. forest land, 28 percent is government-owned, 30 percent is on farms, 29 percent on non-farmland, and 13 percent on land owned by industry (federal and state governments own approximately 38 percent of all the land in the United States, and more than 90 percent in Alaska). Alaska, which is twice the sixe of Texas, is almost entirely wild and for lorn. Approximately 33 percent of Alaska's total area is forested.

In general, mature, natural forests do not produce as much wood as do young forests, though older stands are usually more varied and impressive than second growth. Since 1900, about 16 million acres of forest land have been set aside for parks and wilderness areas. In September 1972, President Nixon asked Congress to add 35 million acres to the nation's wilderness system.

New tree growth in America probably exceeds tree harvests. But many of the trees are growing unnoticed in cities, parks and along roads where the lumber companies cannot get at them.

Less than 2 percent of United States land is urbanized. A few miles outside of any city-and sometimes within its borders-one can find groves of trees or wilderness. Wilderness is a relative thing. To the American naturalist, John Muir, wilderness meant the entire Sierra range. "Thousands of tired, nerve-shaken, overcivilized people are beginning to find out that going to the mountains is going home; that wildness is necessity; and that mountain parks and reservations are useful not only as fountains of timber and irrigating rivers, but as fountains of life," wrote Muir in 1898. To a city boy "wilderness" may only be an unkempt acre of weeds.

VALUE OF FORESTS

Forests have great value to man in that they provide wood and wood products such as the paper on which this book is printed Forests form a habitat for a variety of animals, but most forest animals are small and have little value for modern man. Trees hold the soil and water in place on hilly slopes and thus prevent erosion. Forests provide areas for outdoor recreation for an increasing number of people who wish to tramp or camp away from home.

A sizable jump in lumber and plywood prices during 1971-72 would seem to indicate that commercial timber is in short supply, though the demand for wood products in 1960 was about the same as it was in 1910. Nevertheless conservationists are attempting to reduce timber production in the national forests and to restrict cutting on industry-owned lands, which hold about 12 percent of the remaining virgin timber.

Sometimes forests make use of hilly and infertile soils that may be too sandy or too acid for crops or grasslands. In rocky or infertile areas, forests provide for their own conservation by creation and use of required chemical substances. These forests can only be protective and any cropping would cause serious erosion and a consequent desert area such as we find in the Mideast. A protective forest once planted can supply only a few utilizable products. However, this type of vegetation is productive in the sense that it prevents or reduces erosion and the washing away of barren debris which could harm productive farmlands situated nearby at lower levels.

Trees are often beautiful; a few provide fruit and nuts; some give shade and act as wind breaks. But forests cannot provide the food that modern society requires. I love trees, but since I am more concerned with a good life for the many who live in cities, rather than with aesthetic retreats for the elitist few, I cannot wholeheartedly join with those who fight for the preservation of every forest and every tree at any cost.

CREATION AND DESTRUCTION OF FORESTS

Forests kill off the grasses and smaller plants near the ground by shading them from life-giving sunlight. To grass and shrubs, trees are deadly enemies. Forests are destroyed by fire, natural enemies, and man. I will use forests of the United States as an example.

North American Indians extended the range of their main protein supply-the bison-by burning forests. The Indian fires helped create the Great Plains with its rich store of black soil.

Man can help forests grow or watch them die through neglect. In 1941, more than 190,000 forest fires destroyed over 26 million acres of forest. By 1965, man had reduced the number of fires to about 100,000, with a resulting loss of 2.7 million acres. Firebreaks cut through the forests; airplanes, and educational campaigns have helped reduce the ravages of forest fires so that fire now accounts for only 5 percent of all timber mortality. By far the greatest loss of timber, more than 40 percent, is caused by insect pests. Because of recent bans on the use of insecticides, millions of trees have been blighted by insect pests such as the gypsy moth, which has ravaged the Eastern woodlands, and insects have run rampant in Swedish forests since DDT was banned.

In many forests, browsing animals such as deer and elk prevent trees from reproducing. Losses often take the form of reduced growth and deformed and injured trees Seeds may be eaten by mice or other animals so that the forest does not regenerate at all.

Forests are depleted and sometimes destroyed by improper cutting. The cutting may be selective in that only certain mature trees are harvested, or the loggers may resort to clear cutting-the practice of removing all the trees from a forested area.

Studies in western states indicate that the nutrient content of forest soil takes thousands of years to evolve. Dr. Robert Curry, of the University of Montana, found that clear-cutting is often more ecologically disastrous than forest fires, and depletes the soil of nitrogen and other nutrients Two or three crops of timber may be all that can be expected in many clear-cut areas.

Hurlon C. Ray of the Northwest Region Office of the Federal Water Quality Administration has studied the dangers to water supplies posed by clear-cutting. Ray has found that sediment in streams increases 7,000 times in some clear-cut areas, from less than 10 parts per million of sediment under natural conditions to more than 70,000 in improperly logged areas. Sedimentation causes loss of natural stream vegetation and destroys fish habitat many miles downstream from the clear-cut site.

The conservation position on clear-cutting and forest management articulated by Gordon Robinson, Sierra Club staff forester, in the February 1971 Sierra Club Bulletin, makes sense. According to Robinson, good forestry practice has four main characteristics:

> 1) It consists of limiting the cutting of timber to that which can be removed annually in perpetuity.

2) It consists of growing timber on long rotations, generally from one to two hundred years depending on the species and the quality of the soil, but in any case allowing trees to reach full maturity before being cut.

3) It consists of practicing a selection system of cutting wherever this is consistent with the biological requirements of species involved and, where this is not the case, keeping the openings no larger than necessary to meet these requirements.

4) Finally, it consists of taking extreme precautions to protect the soil, our all-important basic resource.

For some time the lumber companies have recognized the conservationist opposition to clear-cutting and have been working on alternatives. A report in The New York Times of September 7, 1971, describes a new timber utilization method developed by U.S. Plywood that should leave cut areas aesthetically attractive and double the amount of wood fiber that could be obtained from first-growth forests. Specialized logging equipment developed by the lumber industry cuts pine trees up to 24 inches in diameter and then bunches the logs for pickup. During the second stage of the operation, a skidder machine moves commercial size trees to a loading area, while moving all other material, except stumps, to a chipper site where a third machine reduces tops, branches, and leaves to chips which are then screened and sized. The pulverized leaf particles are blown back to the ground. In traditional clear-cutting all the litter would have been heaped into piles and then burnt. The new method not only leaves the forest looking more natural, but allows reseeding to proceed two or three years sooner. Selective cutting and new techniques are likely to increase the cost of lumber operations, but are worth the extra cost in the long run.

Conversion of forest land to other uses has contributed to the decline of forests. Cities are essential if we are to create a viable environment for the growing majority of humanity, and of course farms are necessary to feed mankind. However, suburban sprawl and roads for automobiles cause a tremendous and, I believe, needless waste of forest land. One hundred families living on one-quarter acre plots in the suburbs take up 25 acres of land, while these same 100 families could, in my opinion, be better housed in one 25-story apartment house which would take up less than an acre, thus allowing the remaining 24 acres to remain wooded. It is not uncommon to find 300 to 500 dwelling units per acre in communities such as New York and San Francisco. Compared to cities, suburbs are of questionable ecological value, not only because they hog land, but also because they are wasteful of energy and natural resources.

In part forests are a casualty of the way Americans house themselves; too many, the majority of Americans live in single-family houses rather than in high-rise buildings. The average house makes extensive use of wood in its structure, framing and joists, whereas high-rise buildings are normally constructed of steel, brick and concrete, thus obviating the need to cut down forests in order to supply building lumber. The average multi-family dwelling uses less than two-fifths as much wood per family unit as does the single-family house.

Roads devour a tremendous quantity of land in America, and now cover an area equivalent to all of New England south of Maine plus the state of Delaware. One third of Los Angeles and over 50 percent of downtown Washington, D.C., are given over to roads and garages for cars. The area covered by airports and industry is relatively insignificant compared to suburban developments and roadways.

Environments are never destroyed; they are merely changed into other ecological forms. Lakes become swamps; swamps become meadows; meadows become forests. The forests themselves are battle grounds between competing species of trees. A given climate and soil usually support one type of tree best and, left alone, this one species will gradually eliminate competing types, forming a monoculture of say pine, oak, mangrove, or spruce. It is an ecological law that in a given time and place nature (as well as man) drives out weaker species,

244

and one strong type will become dominant. Man has helped nature create monocultures in agriculture and in forestry, over the loud cries of some "naturalists."

But who is to say which ecosystem is best or most "natural?" The western highlands of England and Scotland are covered with bleak moors, treeless, brush-filled expanses, where once Heathcliff roamed. One group of British conservationists would like to plant trees on the moors, but another apparently more powerful group will not hear of it. They want to keep the moors "natural." Natural? This whole area was once thickly forested. Early Britons, looking for wood to build ships and houses, fuel for fires, and space to graze animals, destroyed the original forests and left the land with the bleak ecological regime to which modern Britons have grown emotionally attached. By the 17th Century, England lacked sufficient timber for shipbuilding. Other once-great maritime powers have suffered because their forests had been cut: Portugal, Spain, the city states of North Italy, Rome, Greece and Phonecia. Phonecia lost out as a naval power when the Cedars of Lebanon were no more.

David Brower and his Friends of the Earth argue against any cutting in national forests and state parks such as the Adirondacks preserve in New York. This desire to leave the forests "forever wild" and "virgin" does not take into account the tendency of nature as well as man to destroy and regenerate forests. What could be prettier or more "natural" than the wooded hills of upper New York State and Vermont? When the early settlers arrived on the Eastern woodlands, the first thing they did was chop down the existing virgin forests. The "natural" landscape we admire today consists of the second or third growths of trees pruned by the local inhabitants. Forests are created by disturbances, both natural and man-made. Fires, disease, insects, and harvesting fortunately prune away many old trees and dead wood so that new "virgin" forests can grow. Though antiques may appear charming to antiquarians, most men prefer young virgins to old ones.

MAN AND FORESTS

Primitive man acted as though forests were at best a nuisance and a poor source of food compared to grasslands. Consequently he burned down trees and cleared the forests wherever he could. Today we sometimes hear statements from relatively sophisticated contemporaries about the absolute necessity to mankind of wild woods and trees. What is the real worth of forests to man?

Last summer I spent a few days in the woods around Port Jervis in the Shawanagunk Mountains about 60 miles from New York. The trees thereabouts have been spared because the land is too hilly to farm. Wandering through the woods I was struck by the almost complete absence of food for human beings. I spotted a possum up in a tree, but if I were starving, I doubt if I could have caught it. Gypsy moths, ants, and mosquitos abounded. Not much else.

Forests provide relatively little food for human beings. Rather, trees lock up large quantities of carbonaceous material in trunks and foliage, neither of which are usually edible by man. Only a few trees provide fruit or nuts that can be eaten by human beings. Most forest animals are small: birds, squirrels, snakes, foxes, chipmunks and possum. Deer and other grazers are usually few in number in deep forests, because there is no grass.

In contrast to forests, grasslands provide fodder for large herds of animals such as buffalo, cattle, and (in Africa) zebra. Tropical savannas are covered by a variety of plants that afford nourishment to man and beast. The sun reaches down to the ground to allow plants to flourish. Animals thrive when the trees are few.

Only by converting forests to grass or farmland has man been able to provide enough food for himself. In many ways forests are the enemies of any mass civilization. Recently a Manhattan neighbor of mine told me that she believed that children who grow up without trees around them are psychologically deprived, that there is an inherent need for humans to dwell amidst trees. Consider the Indians of the American Southwest, the Bedouins of Arabia and North Africa, the savanna dwellers of Asia and Africa and the Eskimos of the ice-covered arctic who live in virtually treeless environments. My neighbor's point of view is bandied about by city dwellers who would be unhappy living in paradise.

How many trees does urban man need? The most citified of families-those in New York-have more than one each. Two and a half million trees grace New York parks and another 500,000 trees line the city's streets, and their number increases by about 2,000 per year. The "need" for trees is purely subjective and quite arbitrary. To a dog a fire hydrant will do as well as a tree.

Forests come, forests go. Do we, living in America's cities in the 1970's need them? Is it true that "all of this forest, man included, is in peril," as claimed by the National Parks and Conservation Association (with 52,000 members) in its recruiting pamphlet? Hardly, especially when we realize that forests kill grasslands and lock up carbon in non-nutritious forms. We may well consider forests as growths of not particularly useful weeds, and wild forests as not essential to man's well-being.

In this respect, some comments of philosopher-long-shoreman, Eric Hoffer, are apropos. He writes in First Things, Last Things of his experience in the western woods:

> I spent a good part of my life close to nature as migratory worker, lumberjack and placer miner. Mother Nature was breathing down my neck, so to speak, and I had the feeling that she did not want me around. I was bitten by every sort of insect, and scratched by burrs, foxtails and thorns. My clothes were torn by buckbrush and tangled manzanita. Hard clods pushed against my ribs when I lay down to rest, and grime ate its way into every pore of my body. Everything around me was telling me all the time to roll up and be gone. I was an unwanted intruder. I could never be at home in nature the way trees, flowers, and birds are at home in human habitations, even in the city. I did not feel at ease until my feet touched the paved road...

> Vaguely at first then more distinctly I realized that man is an eternal stranger on this planet. He became a stranger when he cut himself off from the

247

rest of creation and became human. From this incurable strangeness stems our incurable insecurity, our unfulfillable craving for roots, our passion to cover the planet with man-made compounds, our need for the city, a citadel against the encroachment of nature...

Selected city dwellers find pleasure in spending a few days of each year in a wild forest, tramping, camping or watching birds. Many of us with A. E. Houseman will sing:

And since to look at things in bloom
Fifty springs are little room,
About the woodlands I will go
To see the cherry hung with snow.

Wilderness is, however, not a necessary requirement for a happy, productive life, and most citizens assiduously ignore the natural forests within their easy reach. Man's greatest achievement and creations were conceived in cities, and not in lonely deserts or forests. His greater pleasures and future, if he has one, lie in mingling with humanity in bigger and better cities.

The wilderness is not really needed by urban man, why then should he preserve it? For a number of reasons that have nothing to do with the myth that wild forests and deserts are necessities to urban man: to conserve land for future use, and at the same time to prevent the cancerous spread of suburbs, roads and industry. By judicious use of land we can force the creation of efficient and pleasant cities rather than single-dwelling-unit sprawls. Only by reorienting our thinking regarding places to live and ways to travel can we substantially improve the environment and decrease pollution as our population increases and becomes richer.

ENDANGERED SPECIES — MAN'S EXTINCTION OF ANIMALS BESIDES HIMSELF

The dog tried to cross the street
On his little padded feet-

A Ford hit him!!! Squashed him flat-
So for that dog, that was that.

He'll never cross that street again,
He'll cross no streets at all...
He'll gnaw no more on chicken bones,
Nor pee against the wall.

Peter Agnos, c1966 "Mud Pies"

Congressman Edward I. Koch's office in the Federal
Building overlooks the skyscrapers and the rivers of
Lower Manhattan. He appeared somewhat perturbed the
day I visited him and well he might be. The evil forces of
Republican reaction had chopped up his district and thrown
it in with that of Congresswoman Bella Abzug, making a
race likely against the fiery woman. Both Koch and Abzug
are liberal Democrats, but Koch could not at this late date
change his sex to garner the rising tide of women's lib
votes. Hence his melancholy was probably justified as he
slouched his tall frame in an easy chair and bowed his
balding head. As it turned out, Ms. Abzug decided to
challenge the late Congressman William Ryan.

Having just read the titles of environmental laws
sponsored by Koch during the 1970-1972 session of Congress, I said to him, "I notice you've introduced quite a
number of bills to aid our fine feathered, furry and finny
friends " "Yes," he said. "I'm very interested in environment and nature." "Do you have any wild animals in your
district?" I asked. Koch's district consists mainly of high-
rise buildings, apartment houses and office skyscrapers.
It includes part of Greenwich Village and the fancy East
Side once represented by John V. Lindsay. Strange humanoids wander through the streets of Koch's district, but
none has as yet been classified (legally) as endangered
species.

Among the environmental bills introduced by Koch
were H.R. 77, regulating the killing of northern fur seals;
H.R. 4221, protecting wild horses and burros on public
lands; H.R. 6558, protecting ocean mammals from being
pursued, harrased, or killed (the only ocean mammals
Koch's constituents are likely to meet are lifeguards on

Fire Island); H.R. 8099, extending the Endangered Species Conservation Act; and H.R. 10214, setting national policy for wild predatory mammals (in Koch's district the prime predators one is likely to meet are lawyers and stockbrokers from Yale).

"Look," he said, "my constituents are interested in endangered species. So I sponsor or cosponsor these bills. What do I know about fur seals or wild horses? I come from Brooklyn."

"Aren't there more important issues in your district than endangered wild animals? Like housing, crime, employment..."

"Sure. And I'm interested in all of these things. And I've proposed legislation on all of these things...But when that television show-the one showing all those baby seals being clubbed to death by fur trappers-appeared, my office was inundated with calls to save the baby seals with their big, brown eyes. So I sponsored a law to regulate the killing of fur seals...The same with wild horses. Most of my constituents never saw a seal except in Central Park Zoo. But I do what they ask..."

After talking to Koch and to many other people deeply concerned about wild animals, and reading the passionate endearments of walruses, mustangs, et cetera, by the Times editorialists from the zoo on West 43rd Street, the rule governing modern man's interest in wild animals occurred to me, the Seventh Law of Practical Ecology, viz: the average urban citizen's interest in preserving wild animals increases in direct proportion to his distance from the beasts.

S. Dillon Ripley of the Smithsonian Institution recently estimated that a majority of animal species now alive will be extinct by the year 2000... In a bulletin from The Nature Conservancy, one reads, "It is vital that our forests, streams, ponds, and marshes; our remaining prairie lands; our undeveloped seacoasts and offshore islands; our creatures and plants that thrive only in the wild; that all this wealth be protected before the rapidly accelerating urbanization of our countryside leaves us and future generations destitute of all natural creation." I would demur and say that while it is not "vital" to save our

250

undeveloped areas from urbanization, it would be pleasant for the elite woodland minority and wise to do so on limited ecological grounds.

Less than 80 species of mammals have become extinct during the past 200 years. Over 99 percent of all the

The DODO (Didus Ineptus) of the Indian Ocean island of Mauritius; a large, dumb pigeon, could not cope with its changed environment and became extinct about 1681.

species that once lived on earth are now extinct, but only a few species have died out entirely as a result of man's activities. One bird that died out with the help of man was the dodo, a good example of a bird whose time had come.

A large, flightless and clumsy bird, the dodo lived on the island of Mauritius in the Indian Ocean. It had no natural enemies until the Dutch arrived at the close of the 16th Century and developed a taste for dodo stew. Dogs and other domestic animals introduced by the Dutch after 1644 took a liking to dodo eggs as well as dodo meat, and in less than 100 years after its discovery, the dodo became extinct. There is no question that man's environment was unhealthy for the dodo, but is humanity much worse off without the dodo? We can ask the same question about other so-called endangered species.

It is likely that the dodo would have become extinct even if man had not discovered its island retreat. All that would have been required is the landing on the island of a couple of mongooses or dogs or a species of birds better adapted to occupy the dodo's precarious ecological niche. Man and his animals arrived first and helped the dodo join Darwin's long list of species that could not adapt.

Animals may die out for a number of reasons. A species can only survive if its births equal or are greater than its losses in succeeding generations. Many factors may contribute to a greater death rate and diminished births for a given population. I am continually amazed and delighted that so many millions of species of plants and animals coexist on the earth.

Man, or more precisely, some men, have gone out of their way to preserve a number of poorly-adapted birds and animals. The whooping crane was on its way to extinction as early as 1880 because rice farming had chased the shy crane from its winter habitat in Louisiana, and grain farming was preempting its nesting grounds in the northern Great Plains. Hunters prized the bird. Collectors stole their eggs. In 1937 a sanctuary was established in Texas, and in 1954 a refuge was established in the crane's summer habitat south of Canada's Great Slave Lake. From a low of 14 in 1939 the whooper population gradually increased to more than 75 in 1972.

Eyeing a Gnat, Whooping Crane Stands on One Leg.

Almost everyone has his own favorite "endangered" species. A friend of mine recently organized an interesting environmental group called BACL. The prospectus for BACL reads as follows:

<u>Friends of the Bedbug and the Crab Louse</u> (BACL) (A new environmental organization)

Many persons in our society have become dedicated to the preservation of vanishing species of animals (not including man). Some small animals that have always lived in intimate contact with man are in danger of extinction due to the use of DDT and other noxious modern methods. In certain so-called civilized countries, a few of our insect neighbors have been decimated to the point of near extinction by liberal use of chemicals. I refer in particular to the bedbug and the crab louse, cute little insects whose very existence depends on their propinquity to human beings.

Our youths are growing up without the experience of ever having slept with a bedbug or having scratched in their shorthairs for a cute little crab louse. Once-common expressions such as "lous y" and "buggy" are becoming meaningless to our young people.

The Friends of BACL are dedicated to preserving the bedbug and the louse from extinction. Those of you who value diversity in nature, who wish to give future metropolitan humanity the chance to experience intimate contact with our fellow animals, who do not wish to tamper with the original plan of the gods, we urge you to join with us in the Friends of BACL. Send a contribution to help this noble cause.

But my friend need not have feared for the louse. Some long-haired youths are seeing to it that lice are readily spread from unwashed crotch to crotch along with scabies and VD. And bedbugs have reappeared in communes where youths comingle with nature and DDT is banned.

In The Origin of Species (1859) Charles Darwin comments:

> I think it inevitably follows, that as new species in the course of time are formed through natural selection, others will become rare and rarer, and finally extinct. The forms which stand in closest competition with those undergoing modification and improvement will naturally suffer most.

Relatively few species of animals have been extinguished as the result of assault by man. Out of several million species, the dodo, great auk, Carolina parakeet, European wild ibex, and a few other mammals and birds totalling perhaps 80 species have been killed off. Most of the 80 or so species were done in by the introduction of more vigorous cosmopolitan animals that follow man wherever he settles: goats and sheep, rats and mice, cats and dogs. Rats have caused the extinction of at least nine species of rails-birds that lived on isolated islands. Goats, often left on islands by passing seamen to provide meat, stripped the islands of all vegetation and cover for native animals. Hungry dogs kill any animals they catch. Man's introduction of rabbits into Australia in 1859 resulted in a plague of over a billion hungry bunnies that denuded large areas of the continent. In an attempt to kill the rabbits, Australians introduced cats that unfortunately showed a partiality for birds' eggs rather than rabbits . To reduce the cats the Australians next introduced dogs who preferred to let the cats alone, and instead ate the seals on the beaches, then a valuable source of food. In the process of Europeanizing Australia, a number of primitive Australian marsupials lost out to the more adaptable Eurasian mammals.

Over 200,000 plants have been introduced into the U.S. from the rest of the world, including such staples as apples, pears, rice and wheat. What would America be like if we could return it to its "natural" pre-Columbian state? What would Italian cooking be without the tomato; or Ireland, without the potato; or Rumania without its national dish, "mamaliga" (corn meal mush)? The U.S. citizen would have to diet exclusively on these New World foods: sweet potato, peanut, pineapple, squashes, beans, potato, strawberry, corn (maize), tomatoes, red pepper, turkey and guinea pig. He would not have at his disposal old-world crops such as apples, cherries, pears, rice, soy beans, coffee, cucumber, peas, lentils, cotton, barley, wheat, rye, grapes, melons, olives, figs, onions, walnuts, yams, bananas, sugar cane, citrus fruits, black pepper and plums; animals such as dogs, cats, sheep, goats, cattle, pigs, reindeer, chickens, and horses.

Speaking of horses, some few thousand environmentalists have organized to protect the horse in its "natural" habitat. Natural indeed! From Prudhoe Bay to Tierra Del Fuego the American Indian never rode horses until they were introduced by Europeans. Ecologically the horse is un-American. And it is therefore ironical to observe newly hatched ecologists politic to preserve them in the United States despite the fact that there are some eight million horses extant. Perhaps next attempts will be made to preserve the Norwegian rat in our ghettoes and the English starling in our cities.

Along with introduced plants and animals have come insect pests such as the gypsy moth, the Japanese beetle, and the fire ant. Over 400 parasitic and predatory insects have been deliberately introduced into the United States in an attempt to control unwelcome six-legged guests.

It is almost inevitable that man, wherever he goes, will import plants and animals he considers beneficial and, inadvertently, globe-trotting pests, so that in time the animal and plant populations of inhabited zones of the earth will become more or less homogenized. But it is worth recalling that 70 percent of the earth's continents are likely to remain uninhabited for many years to come, so that most native species will continue to survive for many generations. Nature parks can insure that odd species will survive indefinitely.

A life in harmony with nature has attracted most citified men at one time or another. Consider Rene Dubos, who writes for us all, "The humanness of life depends above all on the quality of man's relationships to the rest of creation-to the wind and the stars, to flowers and beasts, to smiling and weeping humanity." I will not question the desirability of intimacy with the wind and the stars, or even with our 3.5 billion fellow weepers, but is it really necessary to have close relationships with beasts and flowers? My family lives in an apartment with three cats and an undetermined number of ants and silverfish. In summer we are joined by flies and occasional roaches. We have no plants, with the exception of a large, misshapen snake plant, because two of the cats eat whatever plant we try to grow. The ecology of our apartment is sufficient to our needs without wild beasts and flowers. Since man lives in and is moving into cities in increasing numbers, it is instructive to look at the trees and beasts in the Empire City.

A BESTIARY OF GOTHAM--WILDLIFE IN NEW YORK CITY

A small patch of greenery, City Hall Park, nestles amid the skyscrapers and canyons of downtown Manhattan. The Audubon Society has recorded 90 different species of birds in this one little park. More than 400 species of birds live and visit the city as a whole, cooing and twittering on the 28,000 acres of parks and preserves, in back yards, and on roofs within the city's 350 square miles. Among the more common city birds, one finds immigrant sparrows and starlings, and native pigeons and gulls.

The fish that abound in the Hudson do not appear to know that some conservationists have declared that the river is an open sewer unfit for fish. Bass, shad, perch, blues, whiting, cod, flounder, porgy, mackerel, mullet and eels are among the hundreds of varieties of fish swimming in city waters. Bass and crabs have been increasing during the past few years.

Over 500,000 dogs live here as pets, while packs of wild dogs roam some areas of the outlying boroughs, and even in Manhattan. Cats are more adaptable to apartments, alleys and markets than are dogs, and though no one knows for sure, there are probably over a million cats in New York. While many city families have no cats, those that have them tend to keep two or more since cats together are happier, cleaner and easier to handle. In addition, a considerable number of cats are kept in stores and eating places to discourage other wild forms of city life, such as cockroaches and mice. Quite a few sleek felines wander around the Fulton Fish market, and through the alleys in back of brownstones.

Almost all of my years have been spent in big cities-Brooklyn, San Francisco and Manhattan. I have always lived in apartments and as a child I always had a longing to own a tree. On trips to the country, I would marvel at the countless trees lining the roadways and think it a great shame that no one was sitting in their shade, that small boys were not climbing or swinging from their branches, that lovers were not hiding in them. The millions of trees within the city somehow did not generate the same feelings. I wanted my own tree.

When I was nearly forty, my wife and I bought a shack on a small sandy island in Great South Bay on Long Island. On our small plot were four stunted evergreens. I cleared away some of the undergrowth and dead branches and hung a rope on one for my boys to swing on. Unfortunately, the trees give me little pleasure since the island is hard to reach, I have no boat, and I am usually busy in the city. However, there is much tree-covered wilderness closer at hand. On a recent summer Sunday, fed up with writing, I decided to commune with nature.

The only nature book I can find around the apartment is Forest Trees of Maine (1961). Nice day. Yellowstone Park is too far away, so I board the Riverside Drive bus at 97th Street heading north, walk westward along Dykman Street to the Hudson River where I join two fishermen on sloping, cracked concrete slab, talk a little as their lines drift in the strong current. "Killed the mackerel this year off Sheepshead Bay. So many, we had to give them away." A few weeks ago one of them had caught a 23-pound bass at 125th Street. Usually they catch eels and perch that dominate the Hudson waters between the George Washington Bridge and Yonkers, but none was biting today. I said adios after an hour or so, and walked into Inwood Park.

Dozen or so families picnicing on the green flatlands along the river. Men of varied shades and languages playing baseball with gusto. A jogger jogs.

Most of Inwood Hill Park is between the Hudson River and Broadway, on a granite mountain covered with woods. I climb the western slope from the river level to the Penn Central railroad tracks, climb over a six-foot wire fence and enter a thicket. Scratch my legs on brambles. No one around.

Duck under low branches, through weeds and around thickets, clamber higher, past old oaks, maples, tulip locust trees. Climb for about 20 minutes. Still no people. Across a cracked asphalt path; across the West Side Highway through deep woods. Still no people.

Up a steep slope to a dirt path. Sit down on a log, perspiring. Only the buzzing of mosquitos and cars disturbs the wilderness. No people. Look up honey locust (Gleditsia Triacanthos) in Forest Trees of Maine. Learn, "It attains a height of 70 feet and a diameter of 20 inches. . It has escaped in the town of Paris." Put book away in amazement.

Have been alone in park now for three-quarters of an hour in Manhattan, the smallest, most densely populated county in the United States. Strange.

Walk northward. Hark. Signs of an Indian, or Boy-Scout camp. A ring of rocks with twigs in the middle. See stout rope hanging from a branch. No one around. I grab the rope and swing under the trees. A heavy-set black man comes over a ridge towards me. This first human I have met in nearly an hour is carrying a small blue satchel and wearing a green undershirt. "Be careful of that rope, it's not hung from a very strong branch." I thank him. He walks on.

I walk further to the highest part of the park and find a sort of meadow. I go through a hole in a wire fence and onto a granite rock overlooking the Hudson. See roads below, playing fields, broad river, boats, Jersey Palisades. A diesel comes by hauling a string of box cars. I sit on the rock. Look at the river. A thin grey-haired man with a little boy sits down in the grass behind me. The man is pointing out sailboats and tugs to the boy. They leave.

Return to the little meadow. See two couples. Descend the steep hill through heavy woods towards Broadway. See two elderly women standing near an old man who is lying on his back with a transistor radio on his stomach. Further down, six long-haired youths of undetermined sex sit and talk. Down, down to fields level with the Harlem River.

Come onto large meadow with a monument stone and plaque, "shorakkopoch." Here once stood a 280-year-old tulip tree which died in 1938. Indian tribe lived around here.

Walk to ship canal connecting Hudson and Harlem Rivers. Ducks, gulls-a lone heron (blue-grey back). Walk along embankment to where tall man is throwing a cage on a line into the water.

"Catch anything?"

"Some crabs. They were gone for about five years, but now they're coming back... Caught about a quart of killies this morning... Crabs are back in Jamaica Bay. Weakfish too. Make good eating. They'd been gone for 20 years."

Walk over to Broadway and take the IRT downtown, still pondering the lack of nature lovers in this bit of public wilderness in Manhattan.

NATURE LOVERS AND FRIENDS OF MANKIND

Some men have always preferred animals to their own kind and have devoted their lives to promoting bird sanctuaries while their fellow humans starved.   In this country an enormous amount of money is collected each year for preservation of wildlife.   A wealthy country can tolerate a certain amount of callousness without much harm to the populace, although some campaigns often lead to anti-human campaigns: The National Audubon Society is working to prevent the export of pesticides to hungry Asian countries because DDT may cause the death of some wild birds.

Even at the end of this century, less than 3 percent of this nation's land will be devoted to urbanization. Over 95 percent will still be farmland or wilderness. Of the earth's total land surface about 11 percent is devoted to crops, about 17 percent is used for pastures, perhaps 2 to 3 percent is covered by cities, towns and roads. The leaves about 70 percent of the globe in wilderness, relatively untouched by human activity. Admittedly, construction of dams, felling of forests, mining, irrigation projects, road building and other human activities have caused considerable changes in the nature of local environments. But it is misleading to assume that all these changes have damaged nature.   River harnessing that prevents erosion, irrigation that brings life to desert areas, artificial lakes that allow fish to thrive where there were none before, are, on the whole, improvements on nature. Most modern men realize that they are part of and dependant on the biosphere and that its protection is vital to their existence.

To have time to enjoy the wilderness fully requires money and leisure or the ability to live on air, and thus the wilderness minority is composed mainly of middle- and upper-class people who do not have to work hard (or at all) for their daily bread.

One may justify wilderness with the arguments that one would like to leave a legacy of nature to following generations, whether they will use it or not. Recreation in the wilds does have psychic importance to some individuals, just as ballet, baseball or books have great psychic value to other minority groups. I do not begrudge the wilderness minority their woods, but simply wish to put their aesthetic demands in perspective.

It is a curious fact that man's peripatetic curiosity (as evidenced in the mania for international safaris) may be the best hope for survival of many bizarre but otherwise useless animals. A number of animals have been saved from extinction in forest parks and preserves far from their native habitats. Tourism combined with curiosity and the wealthy individuals who feel kindly toward animals may well preserve oddities of the animal kingdom.

I, personally, prefer diversity to homogeneity, but I realize that this desire for variety is probably not shared by most of the human race. I would like to argue for preservation of diversity, for the hell of it, and perhaps because some species, before they join the fossil kingdom, may have something important and fascinating to tell us. There is no arguing about tastes, and I hope the tour-promoters and the wealthy wild-animal lovers will persevere in their attempts to save the purple beaked lesser tit and her friends. I wish them success. But though I cheer them from the sidelines, there are more pressing world problems that presently engage my concern.

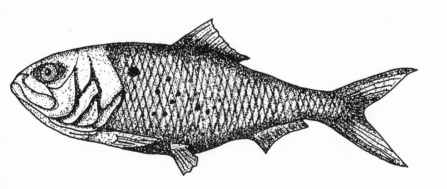

PART THREE: CLEARING THE AIR

I could perhaps, like others, have astonished thee with
strange improbable tales; but, I rather chose to relate
plain matter of fact in the simplest manner and style;
because my principal design was to inform, and not to
amuse thee.

-----from "A Voyage to the
Houyhnhnms"
by Jonathan Swift

16

CLEARING THE AIR

I think I could turn and live with animals,
    they're so placid and self contained,
Stand and look at them long and long.
They do not sweat and whine about their condition,
They do not lie awake in the dark and weep for their sins,
They do not make me sick discussing their duty to God,
Not one is dissatisfied, not one is demented with
    the mania of owning things,
Not one kneels to another, nor to his kind that
    lived thousands of years ago,
Not one is respectable or unhappy over the whole earth.

Walt Whitman

LIFE AND POLLUTION

When launched about five billion years ago, this mud ball was covered with an atmosphere of poisonous gases. Three billion years ago, bacteria, the first primitive forms of living matter, appeared; animals evolved 2.5 billion years later. All living things live and die in response to their existing environment. A species either adapts or dies out. Dinosaurs lasted 160 million years, giving way to mammals about 60 million years ago. That peculiar species, man, has been mucking around for a mere million or so years. Since most forms of life that have evolved are now extinct, man can expect to be replaced at some future date by a species more in tune with the environment of the future. So be it. What we would like to know is how our humanoid tribes can live in harmony with nature in the 20th and perhaps the 21st century. It is foolish to ask for more.

Thus far I have attempted to illuminate a number of flagrant ecological myths and to clarify some questions as to who and what pollutes. Now I would like to consider the more difficult questions: What will the future be like, and what can we do to improve our environment? These questions should be considered in terms of people, that is, what can be done to make the environment better for the greatest number of people. Environmentalists who seem to care less for humans than they do for birds or trees strike me as being peculiar, if not obscene. In considering recipes for the future I concern myself primarily with the five to seven billion human souls who will live on earth in the year 2000, and only secondarily with the feelings of the quaint animal lover who easily ignores suffering children a few miles from his air-conditioned house.

The gods who put us on earth have apparently grown tired of us so that we must fend for ourselves. Though we have not yet fouled our planet sufficiently to endanger our existence as a species, there exists the slim possibility that we Americans, and the other 93 percent of the world avidly copying our technology, will cover our planet with a blanket of contaminants unless rational action is taken. Let us first agree not to jump to hasty and often erroneous environmental conclusions that may lead to rash actions. There is much wisdom in the remark of Philip Handler, President of the National Academy of Sciences:

> My special plea is that we do not, out of a combination of emotional zeal and ecological ignorance, romanticizing about the "good old days" that never were, hastily substitute environmental tragedy for existing environmental deterioration. Let's not replace known devils by insufficiently understood unknown devils.

In this chapter I highlight some of the trends in energy use and population growth, some technological solutions to pollution problems, and the relativity of environmental issues.

THE WORLD ENVIRONMENT—GLOBAL CONSIDERATIONS

Great Britain uses tall smokestacks to dissipate the waste gases from its power plants. As a consequence, British sulfur oxides and particulates have been landing in Scandanavia and mucking up its environment. Noxious effluents put into the Rhine in Germany have killed fish in Holland's part of the river. Air and water currents know no national boundries

Conservation is very much a global concern since air and water pollutants are international wanderers. In May 1969, U Thant, Secretary-General of the United Nations, said:

> I do not wish to seem overdramatic, but I can only conclude from the information that is available to me as Secretary-General, that the Members of the United Nations have perhaps 10 years left in which to subordinate their ancient quarrels and launch a global partnership to curb the arms race, to improve the human environment, to defuse the population explosion and to supply the required momentum to development efforts. If such a global partnership is not forged within the next decade, then I very much fear that the problems I have mentioned will have reached such staggering proportions that they will be beyond our capacity to control.

Maurice Frederick Strong, a Canadian, organized the first United Nations Conference on the Human Environment, which was held in Stockholm in June 1972. Representatives from 130 countries and dozens of international organizations gathered for two weeks to thrash out six major problems: Natural resource management--the environmental aspects of conserving and using animal, plant, and mineral resources; Environmental quality of human habitats throughout the world, and what can be done to improve them; Identification and control of environmental pollutants; How economic development effects the environment, and what policies and plans can be adopted to protect the environment while maintaining

265

economic growth; Social, cultural and educational requirements and effects of pollution; Implications for international
organizations and proposals for positive action.

Some concrete results of the Stockholm Conference
are worth noting:

A few of the lesser developed nations appeared to
recognize the importance of pollution and environmental
problems, and the advisability of dealing with the problems
on an international basis.

The delegates passed a Declaration for the Human
Environment which in effect creates a basis for environmental international law.

More than 100 proposals for action were agreed to by
committees and delegates; for instance, ocean pollution
and climates will be monitored, and vanishing species will
be protected. The conference set up:

1) A governing council of 54 members from 54 nations,
less than half of the nations in the world, but those with
most of its money, power and peoples. The council will
report to the Economic and Social Council of the United
Nations General Assembly.

2) An environmental staff within the U.N. Secretariat
which will deal with international environmental matters
mainly by coordinating the activities of existing agencies.
The staff will be paid by a "$100 million voluntary fund"
of which the U.S. has pledged to provide $40 million.

Strict population control was not voted by the delegates,
nor was limitation of industrial growth in underdeveloped
nations. Third World delegates were content to allow the
richer nations the luxury of industrial stagnation, or vast
expenditures for pollution control projects. Perhaps, as
Margaret Mead said, "This is a revolution in thought fully
comparable to the Copernican revolution...," but it is a
revolution promoted by a small, well-fed minority. The
poorer nations prefer growth with some pollution, to
poverty and stagnation.

No final sollution exists for dealing with international
pollution: there are too many gaps in what we know. Besides, many governments are simply not ready to spend
vast sums on controlling pollution in view of more pressing
national problems

Many U.N. agencies have been concerned with environmental problems for a long time. The World Health Organization has studied population problems, environmental health problems and such issues as use of DDT. The Food and Agricultural Organization has been involved in conservation of soil, forest and animal resources. U.N. Regional Economic Commissions became involved in regional pollution and environmental problems.

The world oceans cover 70 percent of the planet and its waters wash the shores of hundreds of nations. Therefore the U.N. must inevitably deal with ocean pollution issues. An intergovernmental working group on marine pollution has been considering various proposals for the prevention of ocean dumping from ships, though the amount of waste contaminants and oil deliberately dumped into the ocean is only a small fraction of the contaminants reaching the sea from runoff, rain, and natural sources.

Large-scale planning and implementation of major planetary projects will require governmental action and international coordination and cooperation. This viewpoint has been expressed by the distinguished Soviet geophysicist Dr. E.K. Federov, Director of the Soviet Hydro-Meteorological Service. In a 1971 address to the Fifth Congress of World Meterological Organizations, he said:

> It is not difficult to understand that the problem of transforming the climate on a world or regional scale is, by its very nature, an international one, requiring the united efforts and the coordination of the activities of all countries. Ever more rapidly, humanity is approaching the stage in its symbiosis with nature, when it can turn to practical account all the natural resources of the Earth and when, as a result, it will become capable of thinking in terms of natural phenomena on a planetary scale. In other words, man is becoming master of the Earth. It is obviously no accident that this period coincides with our penetration of outer space. It is hardly necessary to prove, that in these circumstances, all

mankind should regard itself as a single whole in relation to the surrounding world. There is no other way.

Federov, and indeed most of the less industrialized non-English-speaking world, still feel that nature can be turned "to practical account," and that man is indeed master of the Earth, a much more positive attitude toward the future than one finds among many persons in the United States and Western Europe.

FORECASTING ENERGY DEMANDS:
THE "ENERGY CRISIS"

Demand for energy, as indicated in the chart opposite, has been growing rapidly in the United States. At the same time the power industry is facing increasingly stringent fuel requirements designed to maintain a clean environment. Coal and oil-fired electric power plants are among the largest sources of air pollutants in the U.S. In the U.S. power plants emit more than 50 percent of all sulfur dioxide and some 25 percent of particulate matter. Most of this was from coal-fired plants. In addition, electric power plants pour heated effluents into waterways, often causing thermal pollution.

Economic well-being and use of energy usually grow together, but there are peculiar inconsistancies in present and projected energy use. In 1971, the average U.S. citizen used some 350 million BTU (British Thermal Units of power) per year. But the average Japanese citizen used approximately 180 million BTU, about half that of a U.S. citizen. What are the major reasons for the great difference in consumption in these two, highly industrialized nations? For one, the Japanese transportation system is much more compact and efficient than the U.S. system, which depends to a great extent on individual automobiles. Also, insulation in smaller Japanese buildings is probably superior to that of U.S. buildings. I recall that when I visited Japan in the 1950's, I was amazed to find houses with double paper walls. A large airspace between the walls acts as an excellent insulator, making it much easier to control the inside temperature and thus conserve energy.

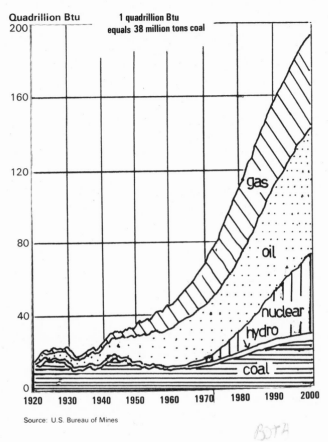

Figure 16-1 Past and Projected U.S. Demand for Energy

	1950	1959	Projected For 1980	Projected For 2000
Population (Millions)	152	204	235	308
Power Consumed Per Person Per Year (Kilowatt-hours)	2000	7000	13,000	26,000

Figure 16-2 AEC Projections of Population and Power Consumption Growth

269

For many years the use by consumers of electrical appliances, electric heating and air-conditioning has been vigorously promoted by power companies, manufacturers of electrical equipment and indirectly by the Consumers Union. As a consequence the quantity of electric power used by individuals has increased in some cases more than double in the past ten years. All electric houses (as compared to homes where gas is used for cooking and gas, oil or coal is used for heating) are large consumers of electrical energy. Electrical energy used for heating purposes is extremely inefficient; first a fossil fuel is used to generate electricity, then the electricity is transmitted to a home, then the consumer converts the electricity back to heat at an overall efficiency of 15 to 25 percent. If the fossil fuel were used directly for its heat, efficiencies of 75 percent or over would be realized.

Thomas E. Browne, principal ecologist of the New York State Public Service Commission, reported that residents of New York City on the average use approximately half as much fuel and power as comparative U.S. suburban residents. One obvious way to save enormous quantities of fuel is to persuade suburbanites to live like New Yorkers.

The projections indicating accelerating power demands are derived from extrapolating growth during the 1950's and 1960's. During the 1960's, the largest user of electricity in this country was the metal fabricating industry. The aluminum industry alone took 38 billion kilowatt hours to produce two million tons of metal.

"Ah," said the projectors in the 1960's, "the use of electricity will increase because aluminum makers will make more aluminum to go into more and more beer and soft-drink cans..." But in 1973 one could observe at least three countertrends: the aluminum industry was lowering prices because of lack of demand; recycling of aluminum cans, while not yet an assured success, was being tried, and could greatly change the whole aluminum demand picture; and some of the nations where bauxite is being mined were building power plants so that they could process their own ore and thus get higher prices on the world market. All of these trends-not forseen in 1960, should drastically cut the demand for electricity in the 1970's and 1980's.

The New York State Society of Professional Engineers made a study of power needs and projections. Its report, issued in the fall of 1971, made the following recommendations:

1) Require the local power company, Consolidated Edison, to distribute waste heat and chilled water to nearby users, instead of dumping the heated or chilled effluents into the environment;

2) Encourage the development of cluster housing with central heating and air-conditioning systems, such as Co-op Village in the Bronx. Large housing developments can make much more efficient use of energy than scattered houses;

3) Encourage the planting and maintenance of trees, which help regenerate the air and shade the streets.

Many of these recommendations are simply common sense substantiated by authoritative data. One looks at them and says "but of course. Why don't we do it?" I believe we will, or our children will.

While the engineers were completing their report, a Sierra Club task force was holding a three-day conference in Johnson, Vermont (January 14, 1972) to see if the club could change its rather negative "all power pollutes" anti-power stance. The final conference report noted that the nation already suffers from a shortage of electricity and faces a possible doubling of demand for electric power in 10 years. The main question then became how to provide adequate power with the least damage to the environment. One major suggestion was that a federal agency be formed to oversee long and short-range energy questions. The proposed agency would include public members who would help determine sites and design of power plants.

As with other studies by environmental groups of the power problem, recommendations were made to slow down the use of electricity. Participants suggested the development of more efficient electrical devices, the construction of better insulated buildings, and a halt to the promotion of electricity by power companies eager to expand their plants and thus their profit base. Again, these suggestions are simply common sense.

Consolidated Edison, by happenstance, turned out to be one of the "good guys" at the conference because they had sponsored a "Save-A-Watt" campaign during the previous summer in New York. The company put money into this campaign mainly because it did not have surplus electricity due to a combination of poor planning, equipment failure, and new-plant holdups by environmentalists. Con Ed's Chairman, Mr. Charles Luce, pointed up the peculiar nature of electricity demand: "When Tiny Tim (an entertainer) got married on the Johnny Carson Show, our load went up by 200,000 kilowatts. It would have been a tragedy if he'd married on a hot summer afternoon"-when all the air-conditioners would have been turned on.

Projections based on trends are often misleading. The trends publicized by the power monopoly are particularly suspect because of their self-serving nature. Electric power company monopolies-or utilities, as they prefer to be called-make a guaranteed profit based on their invested capital. A plant costing $2 billion can generate twice as much profit as a plant worth only $1 billion-never mind if the company or the public actually needs a $2 billion plant rather than a $1 billion plant. Naturally the private investors who run power companies would prefer a $2 billion rather than a $1 billion plant; hence their delight with surveys and scenarios that show an ever increasing need for electric power.

Projections for increased electric power prepared by power companies indicate an increase of some 4 percent per year for the country as a whole, or an increase that would double demand in about 17 years. These projections are probably too high because of the reasons given and because of the "unexpected" decrease in U.S. population growth recorded in 1970, 1971 and 1972. Further, it is likely that electrical appliances will become more efficient, and that many persons will learn that they can do without certain appliances once they have worn out. This has been the case in our family with respect to electric blankets, toasters and electric can openers.

I am skeptical, however, of claims put forth by activist environmentalists that the consumer can save much electricity by skimping. Attempts to reduce power in the home will have little overall effect on energy demand. Consider the report of the Federal Power Commission called The National Power Survey. It concludes that:

> Even if the American people were willing to cut corners in their use of electricity in the home, which remains to be demonstrated, this of itself would not fundamentally after the national demand outlook... 40 percent of the nation's electrical energy is consumed by industry, 18 percent by commercial enterprise, 12 percent by transmission line losses and small power consumers and only 25 percent by residential users.
>
> Thus, with the residential use accounting for only one-quarter of total use, it would take a 10 percent reduction in today's household consumption to achieve a 2.5 percent reduction in today's total consumption.

For the past two years I have been trying to convince my wife that using an electric dishwasher wastes both electricity and money, but she prefers to spend a few extra dollars per week on her dishwasher rather than spend the time washing dishes by hand. Most other consumers, according to Con Ed, want the convenience of air conditioners, washers, dryers and the hundreds of other appliances created by American industry. Convincing them that they should not have appliances is tilting at windmills. How can one tell people that they should not enjoy more comfortable lives?

But industry, prodded by consumers groups and the government, can produce more efficient devices that will use less electricity. More important, industry can use high voltage transmission lines that will reduce transmission losses, make greater use of insulation that will reduce the need for air-conditioning and space heating. Changes in industrial use should lower electrical demands of industry. All told, according to a 1972 federal report-The Potential for

Energy Conservation, which deals with all forms of energy use – it should be possible within 17 years to reduce energy demand by 20 percent.

Economists delight in predicting future growth based on rising trends. It is a pleasant way for econometricians to spend time and it makes their businessmen-clients happy. But such predictions are often quite misleading. Consider the curve below showing the production of horse manure in the United States during the 19th and 20th centuries. Note the buoyant rise in manure from 1800 to 1900. If one projected the curve upward for our present population one would find that we would be buried in the stuff.

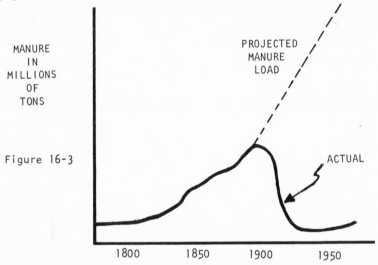

MANURE IN MILLIONS OF TONS

PROJECTED MANURE LOAD

Figure 16-3

ACTUAL

1800   1850   1900   1950

Fortunately the steam engine, electric motor and internal combustion engine have made the work-horse almost obsolete, though there are presently some eight million horses in the United States, used mainly for sport.

As another example of predictions gone awry, consider the office building and space situation in New York City. Clever businessmen advised by expensive economists went on a building boom during the 1950's and 1960's, spurred on by predictions of growing demand for office space. The last great office building boom took place in the 1920's and culminated in the Empire State Building. During the depression that followed the boom, that magnificent edifice was known as the "Empty State Building." But what happened to the projections made during the 1960's? Some companies moved out of the city and business declined in 1970. In 1971 only three million square feet of office space were rented in Manhattan-the lowest amount in a decade. More than 12 million square feet of office space were unoccupied, and half of 1972's 4.4 million square feet of new completions had not been leased. By 1973 there should be some 30 to 40 million square feet of vacant office space. Based on self delusion and rosy projections, New York's shrewd builders have, at great expense, and with lavish use of materials, created enormous steel and glass vacuums. Ambrose Bierce observed the forecasters and noted "The biggest tumble a man can make is to fall over his own bluff."

POPULATION "LIMITS TO GROWTH" AND POLLUTION
From 1650 to 1970 the world's human population increased from half a billion to 3.5 billions. Should the present rates or reproduction and mortality remain constant, allowing an increase of about 2 percent per year, by the end of the century the world's population would double.

Biologist Paul Ehrlich of Stanford University has always predicted war, pestilence and famine as eventual consequences of mankind's growth. "Spaceship Earth is now filled to capacity or beyond and is running out of food," he opined in a lighter moment. Ehrlich and a few other outspoken gloomy prophets have also predicted that crowding, especially in cities, inevitably produces aggressiveness, general disorder and soaring crime rates. Ehrlich made a speech to this effect at the American Association for the Advancement of Science (AAAS) meeting in Chicago in 1971, basing his predictions on work with rats crowded into cages. The charges linking crime and social ills may pertain to Ehrlich's rats, but where humans are concerned the facts indicate a quite different situation. Both crowded London and Holland have remarkably low crime rates, much lower, for instance, than the western and central United States which, though sparsely populated, have high and increasing crime rates.

Certain countries grow faster than others at various times in their history. The United States experienced a 50 percent surge in population during the 30 years from 1940 to 1970. During that period U.S. population grew to a little over 200 million. If this rate of growth were to continue for the next 130 years, the U.S. population would reach a billion before the year 2100. But as a matter of fact, the latest demographic reports indicate that the birth rate has declined appreciably in the United States during 1971 and 1972, to the point where the U.S. population is virtually stable. To the surprise (and chagrin) of those who five years ago were predicting "overpopulation," the nation is approaching zero population growth (ZPG)-2.1 children per child-bearing woman. In March 1973 the Census Bureau published vital statistics that depicted the lowest birth rate in U.S. history-a decline of 10 percent since 1971. The replacement dropped to 2.03 for the 1972 year.

| Projection | Children per woman | Population in millions | | | |
		1970	1980	1990	2000
A	3.350	208.9	249.4	298.1	356.1
B	3.100	207.1	241.9	284.4	331.6
C	2.775	205.5	233.6	268.8	304.4
D	2.450	204.1	226.0	254.0	280.1

Figure 16-4 U.S. Population Projections

Four alternative U.S. population projections made in 1965 by the U.S. Bureau of the Census.
All projections are higher than the real birth rate of 2.11 existing in January 1973.

The major conclusion of <u>Limits to Growth</u> (1972), a report financed by the Club of Rome and written by a team of MIT savants headed by Dennis Meadows, is glum indeed: "If the present growth trends in world population, industrialization pollution, food production, and resource depletion continue unchanged, the limits to growth on this planet will be reached sometime within the next 100 years. The most probable result will be a rather sudden and uncontrollable decline in both population and industrial capacity." Anyone who claims that this conclusion can be reached by reading the slim volume is either a seer or full of beans; the book is one continuous <u>non sequitur</u>.

Charts, graphs, and data supplied by the earnest and well-intentioned authors are grossly insufficient to buttress their case, and are in places misleading. The authors say that the meat of their study--which will show what equations they actually fed into the MIT computer to arrive at their dire predictions--should appear at some later date. At this time there is no possible way for a reader to verify assumptions, equations, and calculations; we are asked to accept their conclusions on faith and many people apparently are only too eager to do so.

As a work of art <u>LTG</u> has quickly become a reflecting pool for those who through ignorance and consequent fear of science and technology are disposed to visions of future doom. (As visual art some of the book's graphs and flow charts might well hang in the Whitney Museum). Mankind has often been beset by visions of apocalypse, millennia, and California sliding into the sea. These beliefs have always been matters of taste and are usually religious in nature. Many bums, astrologers, English majors, and other parasites in our society exist only because science-based technology has been able to produce surplus food and other necessities of life. It is ironic to hear the parasites turn on technology and accuse it of creating most of man's ills. Without modern science the world would support only a relatively meager population, most of it living in sickness and misery.

Looked at purely as a mathematical exercise for a computer, the collapse of <u>LTG's</u> system of equations was inevitable. The MIT savants made certain assumptions about the rate of growth and decline of population, resources, and pollution that preordained the collapse of their model world. About 150 years ago Malthus made somewhat similar mathematical assumptions about rates of population and food growth and arrived at similar conclusions of disaster and collapse. Malthus was wrong in his assumption that population growth is everywhere exponential (it has stabilized and even declined in some areas) and he was wrong in his assumption that food production could not keep up with population growth (more people live and eat well today than at any time in the history of man). One has only to check the increase in life expectancy throughout the world to verify the truth of this statement. In the United States a child born in 1900 had a life expectancy of some 50 years. Today a U S. child can expect to live over 70 years. We appear to be thriving on--or despite--industrial pollution.

A basic assumption of LTG is that pollution is killing people and that the rate at which it is killing people will increase drastically as industrialization increases. But in fact water and air quality have been improving during the past five years in many industrialized areas of the world. The Hudson River is somewhat cleaner now than it was 10 years ago and the crabs are coming back; sulfur dioxide and carbon dioxide concentrations in city air are down from what they were a few years ago. There is little justification for assuming that pollution must grow with technology: replacing cars with modern electric-powered trains will drastically decrease air pollution and solid waste; generating electricity by nuclear plants anchored in the ocean will decrease both air pollution and thermal deterioration of rivers and estuaries.

The aware reader will question the basic methodology of LTG in lumping together data from industrialized nations and countries living at a subsistence, peasant level--countries with life expectancies of 75 years and those with life expectancies of 35 years, countries with natural increases of 3.5 percent and those with decreasing populations, those rich in natural resources and those with virtually none--lumping all of these together and creating the fiction of an average world citizen and an inter-connected world community. The world is not an aggregated community--China with one quarter of humanity could vanish from the earth with virtually no effect on the lives of the rest of the world; India could sink into the southern ocean and hardly anyone in America would be the worse off for it-- though we would have to import tea and swamis from other countries. The world is not a neat statistic that can be fed into a computer, and it is simplistic to think otherwise.

Computers can help us develop a clearer picture of the future than simple intuitive reasoning. But to accept as gospel the results of this admittedly crude and, I feel, misleading study can do more damage than good in the long run.

Believers in LTG's message claim that its authors are taking the long-range "systemic" view of civilization, while their detractors only "understand life as a succession of separate, unrelated single issues." This reminds me of the story about the man who claimed that he handled all the big decisions in his family: he solved the problems of mankind and the universe and allowed his wife to handle family finances, prepare the food, and care for the children. What responsible critics of LTG challenge are the authors' assumptions, arguments, and methods, and it is up to the Club of Rome to show that the LTG study is more than a clever computer exercise in futurology and fear-mongering. Many of the believers project a tone of anti-technology and emotional yearning for a simpler, more stable world. History is strewn with the wreckage of societies that stopped growing, became stable, and then sunk into decay or vanished from the earth.

A person who, without understanding the book's assumptions and techniques, wholeheartedly accepts the conclusion that the end of the world is near will hardly be motivated to attempt to change the present social system, to re-distribute the wealth from the rich to the multiplying poor, to elect officials committed to solving today's problems, rather than those projected for the world of 2070. Who will benefit if activists and revolutionaries turn their thoughts and actions to the fantasy world of the 21st century as outlined in LTG rather than to the very real inequities of 1972?

Rates of change are accelerating, but mainly because scientific findings are being rapidly translated into technological improvements that allow more people to live and to live better than ever before. Man, like other animals, can learn, but he is distinctly different from any other animal in his ability to accumulate knowledge and to teach it to future generations. Knowhow accumulates; men learn to do things better and faster, the old gives way to the new; and foresight lags behind innovation. The question many thoughtful men are asking for the first time in recorded history is: should innovations be instituted? This question is not answered by Limits to Growth.

A fundamental law of computer modeling is:

Garbage In = Garbage Out

In the minds of many laymen, the computer often sanctifies the garbage output though the computer often distorts what little sense there was in the input garbage.

The "fact" that the world's population is growing at about 2 percent per year is a figure that, if accepted without examination, would double the population of the world every 33 years. But most of the large countries of the world have a rate of population growth much smaller than 2 percent.

Japan	1.1%
Soviet Union	1.1%
France	1.0%
U.S.A.	1.0%
China	1.4%

It is only in countries whose people are mainly agricultural and illiterate and who have only recently been introduced to modern medicine and public health techniques, only in these relatively few countries has the population spurted beyond the 2 percent figure. All elements of life and love and nurture on earth are interrelated, complex, and involve many variables and thousands of unknown parameters; precisely for this reason the simple cries of future ecological doom based on population trends are highly dubious.

Most of us live in cities built by men; we see dams, factories, and skyscrapers. The doomsayers cry, "Oh, look what man is doing to our planet! Man is destroying the earth." But few can indeed see the planet earth for the buildings. Humanity has done relatively little to change the earth so far, and even if there were 10 billion of us, man will still have done little to change the basic nature of the planet earth. Less than 2 percent of the total land of the United States is taken up by urban areas. Most of America is as it was when Columbus discovered it: deserts, mountains, forests. In less industrialized countries than the United States, nature has been changed even less in the past 200 years. In many areas, modern man has helped nature by planting grasses and forests, and by preventing floods and erosion. The vast areas served by the TVA were once a flood-ridden wasteland. The Great Plains, where winds once blew away the topsoil in great dust storms, have been transformed by intelligent use of technology into mini-forest belts. New lakes for fish and fowl now dot the countryside. Fertile contoured farmlands hold the soil in place and provide nesting places for wild birds.

After studying the Club of Rome report, Anthony Lewis of The New York Times, wrote, "Fertilizer must be used intensively. . . Fertilizer pollutes nearby streams as it runs off the land and it gradually lowers the fertility of the soil." Quite the opposite is usually the case, for fertilizers have increased the fertility of the soil over vast stretches of the globe. Runoff from properly fertilized and contoured land produces less "pollution," read fertilization, than from many natural areas.

Is the world overpopulated? Perhaps I am biased because I reside with 1.5 million other humans, to say nothing of dogs, cats, mice, rats and pigeons in Manhattan, an island of less than 23 square miles. More than one-third of the island is taken up by parks, train yards and commercial buildings. I do not feel that Manhattan is overpopulated, though we have a population density of some 50,000 residents per square mile. Holland presents a more realistic example of a country that has a high population density, but is a much more desirable place in

281

which to live than other countries with lesser population densities. In Holland, 13 million people live on approximately 13 thousand square miles, a density of 1,000 persons per square mile. Life expectancy in Holland is 74 years at birth; the literacy rate is over 99 percent. "God made the world, but the Dutch made Holland." The rest of the world has much to learn from the way Hollanders use their environment.

The eight countries covering the largest land areas of the planet Earth are, in comparison to Holland, underpopulated, almost unpopulated:

	Area (Square Mile)	Density per Square Mile (1968)	Annual Increase (1963-1970)
Soviet Union	8,647,000	27.5	1.1
Canada	3,851,000	5.4	1.7
China	3,690,000	197.8	1.8
United States	3,614,000	55.7	1.1
Brazil	3,286,000	26.8	3.2
Australia	2,967,000	4.1	2.8
India	1,261,483	415.3	2.5
Argentina	1,072,000	22.0	1.5

Based on what we now know about food and fiber production, the United States could easily support a population of one billion people. We would then have a population density of about 280 persons per square mile. But to live together gracefully, a billion Americans would have to change their life styles.

New Jersey, the "garden state," has some 10,500 farms that produced about 800 million eggs, crops worth $150 million and livestock worth about $100 million in 1971. Chief crops were tomatoes, corn, asparagus, peaches, apples and cranberries. In addition, commercial fishermen netted some $12 million worth of fish and clams. Over 65 percent of New Jersey's land area is devoted to forests and farms. Though a major industrial state, industry and urban areas take up less than 35 percent of New Jersey's land. One might say that New Jersey is practically empty, or at least underdeveloped. The state is strewn with pine forests, swamps, tree-covered mountains in the northwest sector, and the country's juiciest marshland and garbage dump-the Jersey meadows.

Yet New Jersey has the highest population density of any state in the nation, more than 950 people per square mile-a density close to that of Holland. In 1970, 7.2 million New Jerseyans-an increase of 18 percent over 1960-lived in the Garden State. Some 6.4 million of them lived in urban areas, some few hundred thousand lived in slums. But on the average, the New Jerseyan lived better than most of his fellow countrymen.

The United States has only about 56 persons per square mile. Compared to New Jersey (with 950 ppsm) the United States is a virtual desert. And in many ways it actually is: only about 1 percent of U.S. land is devoted to urban developments, where over 70 percent of our population live and work. And Americans are still leaving rural areas for the good life in the cities. By the year 2000, over 90 percent of all Americans will live in urban areas, as they do in New Jersey today.

Now take France (and most Americans would love to). France is self-sufficient, more or less. France has a population density of 240 inhabitants per square mile. This concentration is approximately five times greater than that of the United States as a whole, but of course nowhere as dense as New Jersey.

Paul Ehrlich, the California doomsayer, advocates reducing the U.S. population to 50 million. He feels this would result in a good balance between resources and population. There is no good technical or economic reason why the U.S. could not achieve the population density of France, in which case there would be about one billion Americans. In certain areas blessed by nature, geography and technology, man has already developed life styles based on much higher densities. Americans living in a country of 240 to 300 ppsm would be considerably less crowded than the New Jerseyans of today, who reside in a state that is still 65 percent forest and farm.

GROWTH AND THE FUTURE OF MANKIND

Behold, this dreamer cometh.

Genesis 37:19

Communists and Catholics are united in their opposition to doing much about the growing world population. Many communist governments, on ideological grounds, hold that restriction of growth is not needed in their countries because socialism can increase production to satisfy larger populations. Catholic doctrinaires hold that artificial birth control methods and abortion are against the will of God and thus lead to damnation, which they consider worse than mere starvation.

A point of view held by some doomsayers is that attempts to stem world population increase are too late and will have little effect: the world will witness starvation and plague as predicted by Malthus. However, it seems clear that world-wide catastrophe is not at hand.

The world population is obviously not going to double every 35 years. As literacy, urbanization, and industrialization reach more people, the birth rate will decrease. The recent population stabilization in Japan, Korea, Russia, Western Europe and the United States attest to the ability of nations to control fertility. Of the eight largest nations in the world, nations with over one million square miles each, only India has a population density of over 200 persons per square mile, and only India and Brazil have population increases of over 2 percent per year. These eight nations control more of the world's livable surface than all the other nations of the world combined.

My second reason for viewing increasing world populations with less than alarm is that from 1950 to 1973 the food supply of the world has been increasing at a faster rate than has world population. This is the result of the green revolution: the use of fertilizers, machinery, pesticides, irrigation, improved species of crops, in short scientific agriculture. I have been intimately involved in aquacultural projects and I believe that fish and bivalve farming easily could triple the amount of protein presently harvested from lakes and oceans. Protein can also be made from oil, sewage wastes, and wood. Scientists have developed techniques for converting petroleum into animal food. Any nation with sufficient capital, technology and a scientific philosophy can feed itself.

Of the 13 nations with populations of over 50 million, only India, Indonesia, Bangladesh, Brazil and Nigeria have rates of population increase of over 2 percent. These rates are almost certain to decrease as literacy and urbanization spread. On the other hand the benefits of the green revolution are also diffusing to the lesser developed countries of the world. Therefore it is difficult to become overly alarmed by predictions of world-wide catastrophe. India is experiencing a severe drought in 1973. Unless India and a handful of other nations take more positive actions to reduce their birth rates, and apply technology to food production, they are likely to experience hunger and perhaps starvation in times of drought or natural catastrophe.

The vast majority of the world's people have little to fear from two of the four horsemen of the apocalypse. Hunger has been staved off and pestilence has been checked, but the pale horse with Death riding on its back will always be with us; and the Horseman of War may yet decimate mankind.

> The best authorities are unanimous in saying that a war with hydrogen bombs is quite likely to put an end to the human race... There will be universal death -- sudden only for a fortunate minority, but for the majority a slow torture of disease and disintegration...

So spoke Bertrand Russell long before the instant ecologists turned the concern of the affluent West from the basic survival problems of our time-how to keep peace among the nations; how to save our children from the scourge of war.

17

SOLUTIONS AND NON-SOLUTIONS TO ENVIRONMENTAL PROBLEMS

> The contemplation of things as they are
> without error or confusion
> without substitution or imposture
> is in itself a nobler thing
> than a whole harvest of inventions.
>
> Francis Bacon

Waste results naturally from life: all animals destroy part of their environment in order to exist; all excrete wastes which can poison themselves or other organisms. Even inanimate nature creates and spreads pollutants at random through erruptions of volcanos, erosion of soil by rain and runoff, shifts in climate and other natural catastrophes discussed in Chapter 2.

Since 1972 when Silent Spring jolted the sensibilities of the affluent Western world, an industry and entrenched bureaucracy has sprouted in the United States to study (sometimes to death) and combat environmental deterioration. The popular press has had a field day disseminating "green journalism," entertaining the public with scenarios of ecological disaster and environmental doom. Much of what has been written on the subject is hogwash, as pointed out in detail in Part II of this book.

But some environmental problems are real and serious. Fortunately technologists are grappling successfully with most of them. By wise use of advanced technology, modern man can make his industrialized cities better and more pleasant places in which to work and live. In this chapter I discuss some technological solutions to environmental problems. However, we all can hope to find long-range solutions to many environmental problems that spring from the way we live in the modern world only by re-ordering our life styles; for example, by changing from automobiles to mass transportation; by residing in large, efficient apartment dwellings rather than in suburban single-dwelling units; by using electrical energy more efficiently; and by forcing industry and municipalities to treat waste effluents at their sources.

RECYCLING WASTES: THE PERFECT SOLUTION?

"Every day: six pounds of garbage for every man, woman and child in America... The ultimate solution to our solid waste problem is not incineration or dumping... We must implement a two-pronged program: to generate less waste and re-use what we can... The second part of the program, recycling, holds the greatest promise... Many citizens groups are investing their energy in collection centers." (From Birds Today, People Tomorrow? a report of the Mayor's Council on the Environment, New York Times, April 18, 1971).

The American economy operates to a large extent on the principal of rapid obsolescence with consequent increase in sales. Items as diverse as cars, single-family houses and nylon stockings are designed with finite life-times so that they will deteriorate quickly. This leads to increased solid waste. Excess and often useless packaging designed primarily to help sell a product rather than to protect it are another factor contributing to the growing waste problem. A third factor in the growing U.S. waste load results from the rising cost of labor required to recycle materials, and therefore the reluctance on the part of industry to reclaim and reuse waste materials.

Though constituting only seven percent of the world's population, Americans use nearly half of the earth's industrial raw materials, and produce enormous amounts of solid wastes. In 1971, solid wastes generated in the United States totaled 4.3 billion tons: 360 million tons were household, municipal, and industrial wastes; 2.3 billion tons were agricultural wastes; and 1.7 billion tons were mineral wastes. Of this total, about 200 million tons-some 5.3 pounds per person per day-are picked up by collection agencies, hauled away and disposed of at a cost of over $4.5 billion per year.

By 1980, unless American life styles change, waste collection will mount to over 340 million tons per year- eight pounds per person per day. America's solid waste load is presently increasing at twice the rate of the population. Each year America throws away 50 billion cans, 27 billion bottles and jars, four million tons of plastic, 7.6 million television sets, seven million vehicles and 30 million tons of paper. The disposal problem is aggravated by widespread and increasing use of disposable containers and other convenience materials that do not easily burn or decay.

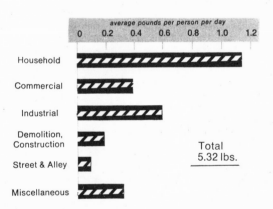

Figure 17-1 Solid Waste Collected in U.S. — 1968

Data: Bureau of Solid Waste Management, U.S. Department of Health, Education and Welfare, October 1968

Figure 17-1 shows that the American solid-waste load consists mainly of household, industrial and commercial trash in that order of importance.

Recycling, the salvaging and reuse of valuable materials in waste, has been jumped on by some environmentalists as the solution to a host of pollution problems. Take out the sulfur from stack gas, they plead, and reuse it; remove the paper from household garbage and recycle it into new products; remove the blight of abandoned automobiles by reusing the two tons of steel in the junk. The recycling concept seems especially appealing because it would solve two problems: it removes vast and growing quantities of solid waste, and at the same time, it conserves minerals, fibers and energy. Since recycling seems to be an unmixed environmental blessing, why isn't it being practiced? As with many other environmental questions there are no simple answers.

RECYCLING AIR AND WATER

Recycling has been tried with varying success on polluted gases, fluids and solids. Certain contaminants, such as sulfur, can be removed from smoke stack effluents. But as yet, the technology combined with the economics does not always provide sufficient incentive for most plant managers to extract sulfur, especially since the price of that yellow element has dropped sharply in recent years. However, sulfur is "produced" as a byproduct in the refining process (or "sweetening") of high-sulfur "sour" natural gas and oil. In fact, byproduct sulfur competes with mined sulfur and in 1971 accounted for more than half of the world's yearly supply.

In a sense, water purification by sewage treatment plants is a form of recycling since the same water may be reused. Signs in certain Chicago pissoirs read, "Remember to Flush, St. Louis Needs Drinking Water." St. Louis water, taken from the Mississippi, may indeed have passed through the gullets of many persons living upstream, to say nothing of animals. Still, the recycled diluted water, purified by natural biological processes, is of good quality. Tertiary water treatment plants theoretically would purify sewage directly, converting it into drinking water without dilution. The problem of health hazards from viruses in the waste effluent remains to plague advocates of complete tertiary recycling. Tertiary treatment is also extremely expensive. To remove all nutrients (phosphates, nitrates, et cetera) and heavy metals (lead, mercury, cadmium, et cetera) could easily triple the cost of sewage treatment plants.

Because of its expense, tertiary treatment can only be justified in extremely water-hungry areas. Tertiary treatment and recycling of wastewater may eventually be built in water-poor areas such as Houston, Los Angeles, Eastern Long Island and some balmy islands in the horse latitudes.

Sewage sludge, containing most of the solids in wastewater sewage, consists of about 95 percent water and 5 percent solids. It is a byproduct of the secondary wastewater treatment process. Most communities pay to have it dumped in the ocean, lagooned or buried. But sludge can be used as a fertilizing and soil conditioning agent. For the past five years, I have been trying to persuade various seaside communities and government agencies of the wisdom of using sewage sludge (now being dumped into the ocean) to build up protective beach dunes along the Atlantic littoral. Sludge mixed with sand makes excellent grass-growing soil; New York City has created a golf course from it. But the New York City Parks Department now claims that it is cheaper to truck in topsoil from surrounding areas rather than to work with the smelly, 95 percent liquid ooze that it could get for nothing from another city agency. Chicago is now experimenting with use of waste sanitary sludge for rejuvenation of farmlands south of the city. Sludge has been used in many parts of the world as a fertilizer and only a certain fastidiousness connected with human waste and economics prevent sludge from being more widely utilized.

Used lubricating and crankcase oil is another fluid that could be recycled, and yet in our society today it is almost valueless and constitutes, instead, a pollutant that often finds its way into our rivers, lakes and oceans. The average motor vehicle uses about five gallons of lubricating oil each year, so that the 100 million American cars and trucks dump 500 million gallons of waste oil each year. Where does it go? A small percentage of the oil is refined and reused. About 10 years ago, legislation was enacted to protect the consumer. It required that all recycled motor oil be labeled. This law put many of the smaller oil scavengers out of business, thus increasing the amount of waste oil throughout the country. Some of the international oil companies have been experimenting with waste oil as an industrial fuel. It unfortunately contains contaminants, such as lead which can foul oil burners, but these problems can be overcome. If the American motorist were forced to trade in his used oil for clean lubricating oil, and a system of pickup and refining were instituted, the waste oil problem would be practically eliminated. Presently economics, the lethergy of the oil companies, and the American tendency to let things slide have resulted in inaction and needless pollution.

SOLID WASTES

An often expressed hope of environmentalists and can producers is that recycling will alleviate the solid-waste problem, and we see well-advertised campaigns mounted to induce consumers to segregate paper and aluminum and glass from their garbage. The garbage problem in America is indeed enormous and has fostered a $4.5 billion industry, euphemistically called "solid waste management". Suburban Americans generate about a ton a year per person of solid waste-over five pounds per person per day. New York City collects over 20,000 tons of refuse each day, or about three pounds for each New Yorker. Households and commercial establishments generate approximately 200 million tons of refuse each year. Residential solid wastes are increasing at about 4 percent per year. They are presently composed of:

Paper products (40-50%)	Increasing trends due to packaging
Food wastes, garbage (12-18%)	Decreasing, since more food is being sold in processed form
Glass and ceramics (8-10%)	
Metals (8-9%)	Probably decreasing as metals are replaced by plastics
Plastics and rubber (3%)	Increasing rapidly
Garden refuse (8-10%)	In suburbs only; central city garden wastes are insignificant

Textiles (3%)	Becoming more plasticized
Wood (2-3%)	Decreasing
Rock, dirt, ash (2-4%)	Decreasing, since coal is being replaced by oil and gas

Figure 17-2 Range of Compositions of Household Solid Wastes:

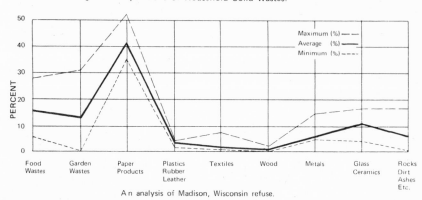

An analysis of Madison, Wisconsin refuse.

THE RECYCLING BUSINESS

Many environmentalists propose recycling as if it were a new solution. Actually, it is as old as the wheel. Most Americans are not aware of the tremendous volume of industrial recycling that presently takes place United States recycling primarily involves industrial rather than domestic wastes, but the latter problem is probably more serious and is receiving increasing attention. The American secondary materials industry includes approximately 9,000 firms with total yearly sales of from $6 billion to $7 billion. Industry recycles a large amount of materials while they are still in production. Scrap aluminum in metal working plants is collected and sold back to the smelter. Rags are collected in a shirt factory and sold

292

to processors. Plastic scraps are gathered in toy factories and sold back to plastic processors.

The largest quantity by weight of solid municipal waste is paper and paper products. In 1969, the paper industry recycled 11.4 million tons of paper, or 20 percent of the total paper consumed. This is the equivalent of 200 million trees, a forest area covering New Hampshire and Vermont combined (14 million acres). In some other parts of the world about 40 percent of all paper wastes is recycled, and there is no technical reason why more than 45 percent of all paper could not be reused in this country. And yet, the percentage of recycled paper has actually <u>declined</u> from 1960 to 1970, due to economic and psychological factors. (A slight increase was recorded in 1971 over 1970). Many purchasers prefer to use paper made from virgin stock rather than recycled paper, although in most cases there is no discernable difference.

Unreasonable policies of the federal government tend to dampen enthusiasm for recycled paper. The government had for many years prohibited the purchase of reclaimed paper in most of its purchase specifications. The government set freight rates that discriminate against movements of bulk waste compared to virgin bulk. The government also provides a 17 percent depletion allowance to the paper companies for their forest-cutting equipment; in essence subsidizing the cutting of trees rather than the reuse of paper products.

If industry and governments were to demand recycled paper, there is no question that its use would grow. In 1971, the New York Chamber of Commerce, through its Special Committee on Environment, surveyed 61 of its corporate members to find out whether or not companies were using recycled paper, and if not, why not. The committee found that most companies have not yet tried recycled paper, though 95 percent of the firms that have used it rated it "excellent," "good" or "fair." One large firm could find no difference between virgin and recycled paper. Of the firms surveyed, 63 percent said that they had not used recycled paper for the following reasons: the paper is limited in supply; it has a "gray" look-the result of insufficient bleaching of the print from the used paper; it lacks the "feel" of virgin paper; its strength is somewhat less than that of virgin stock; or its opacity is sometimes off as is its color consistency. But several companies were simply discouraged by the fact that recycled paper costs more than virgin paper. Most of the above problems can be overcome with increased use and distribution, though now they remain serious barriers to increased industry use of recycled paper. The New York Chamber of Commerce has devoted considerable effort to informing the business community of the desirability of using recycled paper, and as a result a number of substantial firms (Chase Manhattan Bank, American Telephone and Telegraph, Coca Cola, and others) have taken steps to increase purchase of used-fiber paper.

Recently New York City has changed its purchase specifications so as to encourage the reuse of paper and thus cut down on the overall solid-waste problem. These actions and the stated intention of the federal government to use more recycled material should improve the paper recycling picture.

More than three million tons of aluminum, silver, copper and other nonferrous metals are recycled annually in the United States. The amount of copper being reprocessed rose from 42 percent in 1966 to over 50 percent (one million tons) in 1970, and should continue to rise as copper prices rise. Silver used mainly in the photographic industry is largely recovered and reused. The recent rise in silver prices has encouraged improvement in recovery techniques. Conversely a softening in silver prices may dilute this recycling effort. About 35 percent of the silver used in the United States comes from secondary sources.

The reuse of aluminum has also been rising and now amounts to about 25 percent of total U.S. production. Aluminum companies have been riding the ecological-involvement wave by calling attention to their recycling efforts. During 1971, key executives of the major can companies announced the opening of 200 recycling centers. Some of them closed in 1972.

In 1969, more than 36 million tons of iron and steel were recycled by steel companies. However, the percentage of reused steel is likely to decline, since the new oxygen-fired steel furnaces require less scrap iron than the older open-hearth processes. About 30 percent or 360 thousand tons of textiles are reused each year, mainly as rags. Most of the reused material comes from textile and clothing factories.

Reclaimed rubber constitutes about 12 percent of total consumption of new rubber. But anyone who has traveled in the U.S has seen the mountains of old tires that have little prospect of reuse for their rubber value. Why not use old rubber tires for paving roads? Since the tires are free, presumably all one has to do is grind them up and mix them with an aggregate to form an inexpensive roadbed. This sounds attractive and was tried by the Marion County Public Works Department, near Salem, Oregon, in June 1971. Gravel aggregate, which is usually used in road building, costs $3.00 per cubic yard. Aggregate made from free, old car tires after grinding, mixing, shipping, debeading, and so forth, came to $20 per cubic yard. This is an unfortunately typical case of recycling "free" waste material. Tires have been converted into artificial reefs and placed on the continental shelf, where they attract and shelter fish. A small number of old tires have been put to use as swings and jungle gyms in playgrounds.

Plastics, which represent 2 to 3 percent of municipal solid waste, present a particularly difficult recycling problem-especially when they are mixed with other materials, as they usually are. After separation and cleaning, their salvage value is usually lower than new material. Polypropylene and polyethylene account for 55 percent of all plastics in solid wastes. They burn well in incinerators, unlike polyvinyl chloride, which may emit poisonous gases when burned.

Plastics are a fast growing bulk material in municipal wastes. Plastics packaging should grow at an annual rate of 10 percent during this decade, compared with a 5 percent growth rate for all packaging. Approximately four billion pounds of plastic were used in packaging in 1970. The figure should increase by 1980 to about 7.5 billion pounds. Because of their inertness, most plastics make a good landfill component.   Plastics derived from crude oil, coal and natural gas have a high thermal content when burned.

HEALTH HAZARDS IN RECYCLING MUNICIPAL WASTES

To avoid fecal-borne diseases municipal wastes must be picked up quickly and in a sanitary manner. Many infectious diseases such as dysentery grow rapidly in organic wastes.   Rats and mice, mosquitoes, flies and roaches-all of which spread disease-live in unsanitary garbage.   Insects are responsible for "at least two million cases annually in the United States" of food-borne infections, according to Dr. George Kupchik of the American Public Health Association.

During 1970 and 1971 the ecomania induced by the press and TV inspired many well-meaning citizens to engage in recycling expeditions through garbage for aluminum cans, paper, glass bottles, and so forth. Dr. Kupchik comments:

> The hazards of cuts and injuries from sharp metal cans and glass, exploding aerosol cans, punctures from discarded hypodermic needles, the potential for contamination with infected fecal and respiratory wastes are obvious. Would anyone think of rummaging in a bin containing hospital wastes? Yet as you doctors know, there are more sick people at home on any day than there are in all the hospitals and their wastes are put out for collection with the others.   And wastes from doctors' offices are not collected separately either.   Scavenging from waste receptacles is condemned, quite properly, in every civilized country.

The waste materials collected by these enthusiastic volunteers were then brought to collection centers where they were sorted, crushed, and tied as required.   The operations were conducted in cellars, churches, empty lots and stores often under questionable sanitary conditions. In my Manhattan neighborhood, at one particular recycling center started in 1970 during the height of the craze, 100 volunteers lugged five to nine tons per week of bottles, cans, and aluminum plates, to a church basement at

Broadway and 86th Street. The volunteers worked for an average of eight hours per week, contributing 800 man (or child) hours of labor per week to the effort. By the fall of 1971 this effort had disbanded along with the six other recycling centers on the West Side. The volunteers found recycling to be a hard, dirty and unrewarding business. Recyclers received $20 per ton for glass and $200 per ton for aluminum (45,000 cans). The $7 per ton offered for waste paper made any collection effort ludicrous. Paper also would have created a fire hazard. The weekly income of $100-$200 (generated at a labor cost of 12 to 25 cents per hour) was donated to the Boy Scouts. Some recycling centers operated as community projects and run by committed volunteers have proved more lasting. But do-it-yourself recycling doesn't seem to be a solid solution to the solid waste program, and it can create other environmental hazards.

OTHER USES FOR MUNICIPAL WASTES

Burning garbage for its heat value is a form of recycling that has been practiced for many years abroad, but not in the United States. Municipal wastes have a heat value about one-third that of coal, some 1.5 thermal kilowatts per pound. For many years Paris has been burning its garbage to produce steam for heating and electric power generation. If all the garbage in the United States were burned for fuel it would produce less than 10 percent of the required energy for electrical consumption, but it would greatly alleviate the problem of what to do with solid waste. And of course burning produces air pollution-pollutants from certain chlorinated plastics being particularly noxious. However, progress is being made in developing cleaner, improved burners, but nothing is simple when dealing with environmental matters.

Municipal incinerators are prone to a host of peculiar problems. During the summer grass sifts through the grates and adds chlorine and nitrogen gases to the stack; auto tires create a smoke and odor problem; watermelon rinds and wet garbage may simply sear on the outside and later stink as they decompose; aluminum melts and gums up the grates; live amunition, aerosol cans and oils can explode.

In August 1971, Jerome Kretchmer, then New York City's environmental protection administrator, said that the city was abandoning its plans for an incinerator that was to be built and operated jointly under one roof with the Red Hook Water Pollution Control Plan in Brooklyn. The $200-million incinerator, with a 6,000-ton-per-day capacity, would have been the largest in the world. Kretchmer said, "in an intensive rethinking we realized that this plant would be too vast in size, too expensive and too great a concentrated source of air pollution..." Despite air pollution control devices, the plant would emit more than 3,000 tons of particulates each year in an area that could ill use that much pollution. In 1967, proponents of the plant had claimed it would emit only 365 tons of flyash per year.

Japanese recyclers developed a system of compacting garbage into bricks, covering them with asphalt or other impervious materials and using the bricks to construct buildings. Recyclers around the world said that it seemed like an excellent application. Some houses were built. During 1971 some of the garbage-filled bricks began to expand and explode, and the recycled houses came tumbling down. Nature in the form of bacteria decomposed the biodegradable components in the bricks, a reaction that gives off gases such as methane and sulfur dioxide. The gases expanded and exploded the bricks.

THE FRANKLIN COMPLETE GARBAGE--RECLAIMING PROJECT

Franklin, Ohio, a town of 10,000, calls itself "Environmental City." It is the first community in the United States to adopt a system for complete reuse of municipal solid waste. Franklin uses a newly developed recycling plant for municipal solid wastes that takes unsorted household garbage, as collected by refuse trucks, and automatically processes it to reclaim glass, metals and paper-making fibers. Using equipment and techniques adapted from the paper industry, Franklin's recycling method eliminates the need to separate household refuse prior to pickup.

298

Hard, bulky items such as bed springs, radiators and engine blocks are first separated from the garbage. Raw garbage from municipal collection trucks is then fed by conveyor into the water-filled pulping vessel of the system, where powerful rotors beat metal containers into compact balls, pound glass into uniform particles, and reduce paper and cardboard to component fibers. The pulverized trash can then be piped as a slurry containing $3\frac{1}{2}$ percent solids.

Pulping equipment adapted from industry chops materials such as plastics, foils, ceramics and crockery into small uniform pieces. The system uses devices called "junkers" to magnetically remove the compacted ferrous metal, which amounts to about five tons a day. A liquid cyclone centrifuge removes sand, dirt and glass cullet: the plant will put out about three tons per day of glass. Paper makes up about 50 percent of a typical load of municipal garbage: the Franklin system reclaims about 10 tons per day of pulp. Nonrecoverable organic residue from the system is burned completely in a fluid bed reactor that creates no air pollution problem. The inert high-density materials left after all useful or combustible substances are removed can be used for land fill. Unlike garbage land fill, the Franklin fill can be used almost immediately on building sites, instead of waiting the ten years required or a sanitary land fill to settle and deputrify.

The net cost to Franklin, not counting the government subsidy for the experiment, is about $6 per ton of waste. The real cost is much more, and would increase for towns situated further away from potential users of waste iron, aluminum, glass and paper. Larger plants processing over 1,000 tons per day could probably reduce operating costs. The Franklin experiment is a very small step in the right direction.

RECYCLING IN THE FUTURE

Despite lack of government money, a tremendous amount of research on recycling is under way, spurred on by the cry of environmentalists, the disposal problem created by the mounting waste, and the decrease in natural resources. Papers given at the Third Mineral Waste Utilization Symposium (March 1972, Chicago) indicate the trends in the area of reusing industrial wastes, scrap metal, mining wastes and municipal refuse. About 50 papers were presented under four headings: utilization of industrial wastes, mining and milling, scrap metal wastes, and urban refuse.

In perhaps 15 years, if the government acts forcefully, salvage and reuse will be an integral part of most municipal solid waste programs. Alternatives to recycling are incineration and land fill. Incineration produces air pollution and is extremely costly-$8 or more for each ton of refuse burned in New York City. A long-planned incinerator for the Brooklyn Navy Yard site, which was to handle one-third of New York City's solid waste, was cancelled in 1971 because of concern over air pollution and rising costs. Sanitary land fill works well if cheap land is available close to the generated garbage.

New technology is required to deal with the growing solid waste problem. Though recycling should be encouraged when feasible to reduce the burden of solid waste as well as to conserve national resources, we should be aware that this will cost money and effort and that it is only a partial solution to some environmental problems.

Dr. Gordon MacDonald, of President Nixon's Council on Environmental Quality, said in 1971 that recycling was the "key to the successful solution of our solid waste program" and that the government was now involved in a comprehensive study regarding ways to make it work. But although Congress authorized $463 billion for a three-year solid-waste program in the Resource Recovery Act of 1970, the President had requested only $19 million in actual spending for 1971. Spread over the country, $19 million is hardly likely to make a dent in the problem. Congressional authorization for the Office of Solid Waste Management Programs for 1973 is $216 million. The actual Nixon Administration budget request for 1973 is only $23.3 million. In short, the Administration does not intend to spend the money that is, by law, available for the purpose of enhancing our environment. On meeting Dr. MacDonald in New York on March 9, 1972, I asked him why the Nixon administration was not using the money available to it.

"There's nothing to spend it on," he said. "It would be wasted."

At present, the economic rationale for recycling municipal wastes simply does not exist. At $5 per hour and up for garbage collectors, pardon, "solid waste managers," it does not make sense to pick out cans and bottles from the nation's garbage piles. And at least 50 percent of municipal wastes are simply nonsalvageable. Realistically, then, some recycling is now in progress, but the economics and the ecology do not now favor recycling more than a fraction of the country's municipal solid wastes. In the future, as raw material sources are used up, rising prices for some primary products will undoubtedly make recycling more economically attractive. It may be another 10 to 15 years before recycling, given the full government impetus that it does not now have, may provide a substantial solution to the mounting garbage problem.

NEW TOWNS: CITIES OR SPRAWL?

The city is the key to the future. Some planners have suggested that the megalopoli of New York and London are doomed and will be replaced by new towns surrounded by green belts. Others, including myself, argue that larger cities are ecologically superior habitats for future mankind than suburbs or small towns. But there are a few environmental problems that will have to be solved before large cities are made appetizing to all. Prime among the problems is transportation.

"The city does not belong to the automobile," said Premier William Davis of Ontario as he led a fight to stop the construction of a $143 million expressway into Toronto. Some $74 million had already been spent when Premier Davis, with the backing of the Provincial Ontario Cabinet, put a stop to construction. The modern city does not need and can ill afford the automobile when public transportation is adequate.

Cars are simply not acceptable means of mass transportation in genuine cities such as New York, Boston, and San Francisco. Mock cities such as Los Angeles and Houston, which are really clusters of suburbs with baseball stadiums, could not exist without cars. In fact, there is a synergistic relationship between cars and suburbs: each vice feeds on the other.

In Manhattan, (the most reasonable place for an environmentalist to live), our 1.5 million residents own only 200,000 cars. By way of contrast, the 7 million people in Los Angeles own and use 4 million cars, 5 times as many vehicles per person.

I and most Gothamites view the automobile as a generator of over 70 percent (by weight) of our air pollution, a creator of unwanted noise, a device which smells up our streets, clogs our traffic, and kills our pedestrians-in short, an unnecessary pestilential nuisance. When I tell younger people from the suburbs that I do not own a car, they usually assume that I am either too poor or that I have had my license lifted for reckless driving. But I and most of my relatively prosperous Manhattan friends use public transportation by choice. If we must go to the hinterlands we can always rent cars.

Tinkering with auto engines does not appear to be a solution to the overall environmental problems created by unrestricted auto travel. If the future of the United States, where over 80 percent of the population will live in metropolitan areas, is to contain less noise, air pollution, dirt, death, and other environmental hassles, then mock cities must organize themselves along the lines suggested by Manhattan.

Many cities throughout the world are taking steps to keep cars away from the central districts, and if possible to confine them to the suburbs, where the sprawl of single-family houses make them a necessity. Campus-type office parks growing like weeds in some suburbs take up to 20 times as much land as does an equivalent city office building. Corporations moving to the suburbs defend their actions on the grounds of cost and employee convenience, but it is usually for the convenience of the chairman of the board that these environmentally unwise moves are made. Many of the same corporations, needlessly destroying forests and farms, advertise their environmental concern in full-page, four-color displays.

Increased pollution in America derives largely from urban sprawl and automobile use. In order to reverse this trend there must be a fundamental shift in values regarding the desirability of living in multifamily homes in cities as opposed to living in single-family houses in the suburbs. Man has the technology to build durable, efficient 100-story buildings. The Empire State Building (1,250 feet, 102 stories), which opened in 1931, has over two million feet of usable floor space and its base takes up less than two acres. Near my office in downtown Manhattan, the two 110-story buildings of the World Trade Center are nearing completion. These buildings will hold over eight million square feet of usuable space. Since the average person requires 250 square feet of floor space to live comfortably, the World Trade Center could comfortably house over 30,000 persons. The 110-story, 1,450-foot-high Sears Tower is under construction in Chicago. Designer, Fazlur Kahn, predicts that 100-story buildings will become commonplace within the next 10 years. This is environmentally desirable since skyscrapers use land efficiently and create an attractive urban environment. We know how to conserve land by building high, but it is still necessary to bring about the proper socioeconomic climate that will induce more people to live in high-rise apartments rather than in sprawled developments.

The construction of new towns surrounded by green belts has been advanced as a solution to many problems of pollution and population growth. At least the current passion still is strong in America, though it has waned somewhat in Europe. The detached Utopias depicted to prospective buyers are often concocted of nostalgia for the American small town and an attempt to escape from the problems of the cities.

For the past three decades, Americans have left small towns in droves to congregate in the cities where the action is. As a result, some 60 percent of U.S. counties have lost population consistently for the past 30 years. But the migration of poor farmhand blacks, Puerto Ricans and poor whites to the city is practically over and is now down to a trickle.

Experience in England has shown that the wealthier, more active inhabitants tend to leave new towns after a while. Younger and better educated individuals also tend to leave the smaller towns to migrate to the city and there is no reason to believe that new towns will hold this dynamic segment of the population. Bright youngsters emigrate to seek the higher, more varied quality of life available in big cities.

Material well-being tends to increase with urban size. In a large city one can find stimulation of various kinds; one has the advantage of opportunity and high connectivity; one can interact with diverse institutions and humans. These activities are usually physically impossible in smaller towns.   In many smaller towns, with populations of less than 200,000, one large firm can make or break the economy; a man out of work has no place to go for a job; a bright student has little opportunity to follow a career; a woman has fewer stores in which to shop; transportation is often chaotic; and life has a dull quality that leaves much to be desired.   Suburbs often exude an atmosphere both compartmentalized and rigid.   Both rural and suburban communities are generally characterized by distrust and suspicion among neighbors, violence, pettiness, and withdrawal.

Compared to the suburbs and country, the city has much to offer, as suggested in the sonnet of Peter Agnos, who after having spent three weeks in a rural town 30 miles west of New York, wrote:

As one who's been too long in small-town pent
Near deranged with raucous cricket chirping,
And local bumpkins, local-color burping,
Three weeks have dragged leaving much to relent.
Thank god I'm back amidst the warm hubub,
Loose of the rubber-tire kicking clod,
Again with smoothed city humans I hobnob
And with them 'gainst the human pulse rub.
Nature, too natural is for the birds.
Denuded of a vibrant populace
She grows tasteless and varicose,
And reticent for lack of living words.
In the city lies gathered man's great art and song!
In the city humanity's blood flows headlong.

In America today, timid middle-class families are fleeing to the suburbs where, once settled, they prevent the erection of high-rise buildings.   And, they attempt to rezone the area so that no "undesirables" especially, bless the Black, no black can buy or build in the area. While Nassau house plots average about one-quarter acre, further from the city in Suffolk County, plots average one-half acre.   The usual colonization cycle starts when a builder-promoter puts up 30 or so houses in a "development" on one-third acre plots.   Urban sprawl spreads like a cancer, forcing longer car trips and generating more and more pollution. When middle-class Americans learn to appreciate the advantages of city living, many of our pollution problems will automatically disappear.

SUGGESTIONS FROM MADISON AVENUE

"Daddy, What Did You Do in the War Against Pollution?" asks a determined little girl in an advertisement widely promoted by the Advertising Council supported by packaging money. What can one person do? The ad tells us:

> Lots of things--maybe more than you think. Like cleaning your spark plugs every 1000 miles, using detergents in the recommended amounts, by upgrading incinerators to reduce smoke emissions, by proposing and supporting better waste treatment plants in your town. Yes, and throwing litter in a basket instead of in the street.
>
> Above all, let's stop shifting the blame. People start pollution. People can stop it. When enough Americans realize this we'll have a fighting chance in the war against pollution.

Though reasonable on the surface, the prescriptions suggested by the Advertising Council of America in its widely distributed advertisement, do not get to the heart of the American pollution issue. The simple solutions suggested by the Advertising Council will have little effect in reversing the increase in pollution. I would replace the key, penultimate paragraph in its ad with the following:

> What can one person do for the cause? Lots of things, maybe more than you think. Like demolishing your one-family suburban house and moving to an apartment dwelling; like getting rid of your cars and walking, bicycling, or taking a train instead of car-riding; like washing your clothes and dishes less frequently and by hand; like insisting to your congressmen that they push for construction of power plants in isolated areas such as islands and hilltops, away from

population centers; like simplifying your life and not buying overpackaged foods or shoddy appliances.

But I would hardly expect the Advertising Council to endorse these "solutions," since they imply that Americans are polluting their atmosphere, water and land by attempting to buy their way into the Council's notion of the "good life." Herein lies one of the main problems in achieving a balance with nature in today's world. The business community and its lackeys in the advertising profession would prefer not to see a diminution in sales of cars, oil, single-family homes with all the junk that goes into them, appliances, processed foods, and so forth. Industry, with much prodding, has expressed an interest in controlling the air and water pollution from factories. But a major portion of potentially harmful pollutants in America comes directly from the spread of people to the suburbs, people living in single-family homes, with gadgets and automobiles. Control of industrial pollutants will not eradicate air and noise pollutants from automobiles and space heating.

SCIENCE AND TECHNOLOGICAL SOLUTIONS

This is the fire that will help the generations to come-if they use it in a sacred manner. But if they do not use it well, the fire will have the power to do them great harm.

Sioux Indians

Popular concern about environmental deterioration has fostered a distrust in science and in scientists; indeed science is blamed by many laymen for the increase in pollution and for a host of other societal problems. But what is science? To the physicist it is a tool for digging and sorting out the laws of nature, a method for finding and verifying reality. To most laymen, science is a magical art practiced by inscrutable and recondite men in instrument-filled laboratories. The lay environmentalist's superstition about, and reaction against true science flowers through ignorance, an ignorance fostered in large measure by the unwillingness or inability of scientists to communicate easily with the public. The majority of people has replaced its former faith in unseen gods with a faith in science, a faith as blind as the former was. As bygone peoples turned on their gods when things went awry, so do today's populi turn on science and scientists when difficult problems beset society.

During the past hundred years Americans accepted science in a profoundly religious spirit as exemplified by Walt Whitman's poem Song of Myself:

> Hurrah for positive science!   Long live exact demon-
>   stration!
> Fetch stonecrop mixt with cedar and branches of lilac,
> This is the lexicographer, this is the chemist, this
>   made a grammar of the old cartouches,
> These mariners put the ship through dangerous unknown
>   seas,
> This is the geologist, this works with the scapel, and
>   this is the mathematician.
>
> Gentlemen, to your first honors always!
> Your facts are useful, and yet they are not my dwelling,
> I but enter by them to an area of my dwelling.

In the 1940's Bertrand Russell could write about happy scientists and the high status of science in society:

> Of the more highly educated sections of the community, the happiest in the present day are the men of science. Many of the most eminent of them are emotionally simple, and obtain from their work a satisfaction so profound that they can derive pleasure from eating, and even marrying... In their work they are happy because in the modern world science is progressive and powerful, and because its importance is not doubted either by themselves or by laymen. They have therefore no necessity for complex emotions, since the simpler emotions meet with no obstacles. Complexity in emotions is like foam in a river. It is produced by obstacles which break the smoothly flowing current. But so long as the vital energies are

unimpeded, they produce no ripple on the surface, and their strength is not evident to the unobservant.

All the conditions of happiness are realized in the life of the man of science. He has an activity which utilizes his abilities to the full, and he achieves results which appear important not only to himself but to the general public, even when it cannot in the smallest degree understand them. In this he is more fortunate than the artist. When the public cannot understand a picture or a poem, they conclude that it is a bad picture or a bad poem. When they cannot understand the theory of relativity they conclude (rightly) that their education has been insufficient. Consequently Einstein is honored while the best painters are (or at least were) left to starve in garrets, and Einstein is happy while the painters are unhappy. Very few men can be genuinely happy in a life involving continual self-assertion against the skepticism of the mass of mankind, unless they can shut themselves up in a coterie and forget the cold outer world. The man of science has no need of a coterie, since he is thought well of by everybody except his colleagues.

(The Conquest of Happiness, pp. 86-87)

Contrast Russell's statement with the appraisal of science in the world of the 1970's by Murray Gell-Mann, a distinguished physicist and Nobel Laureate.

Some of our most successful institutions are in trouble, under attack, and even despised, sometimes by intellectuals and frequently by educated young people... In our country, in particular, science is in ill repute, together with such gigantic and impressive feats of engineering as the manned flight to the moon...

And that is not all. We are seeing among educated people a resurgence of superstition, extraordinary interest in astrology, palmistry and Velikovsky; there is a surge of rejection of rationality, going far beyond natural science and engineering. In my opinion, some of the adverse reaction to science and engineering and even rationality is understandable.

There are the unfortunate effects of carelessly deployed or carelessly diffused technology.. These effects are interpreted, and quite correctly, as being connected with a kind of narrow rationality, that takes into account in decision-making only things that are very easy to quantify, and sets equal to zero things that are hard to quantify... We see facts and figures marshalled in huge arrays that have failed somehow to include inputs from common or from human values.

How fickle, the attitudes of the public. From adulation and unalloyed admiration of science only a few years ago, the public has changed its viewpoint to suspicion, fear and mistrust. In large measure the current environmentalist campaigns against rational use of materials, nuclear energy, innovations and new developments in general, are reflections of the public trend against scientists and all they represent. College students are continuing their six-year flight away from engineering and science courses. According to an American Council of Education report issued in February 1973, enrollment in the physical sciences has dropped about 50 percent since 1966. Rationality is replaced by near-lunacy on a grand scale.

Some 10,000 "professional" and 175,000 part-time astrologers gaze into the American skies. They cast horoscopes for about 20 million persons during 1969, and in this magical process caused $150 million to change hands, thus proving the value of astrology, at least to those who know how to interpret the meaning of the stars. The majority of people, especially females, have an enormous capacity for believing credulous nonsense and putting their faith in occult systems such as palmistry and astrology. No scientific study has shown that the positions of the planets and the stars control the behavior of human individuals. Nevertheless millions of women (and many men) continue to believe in the astrological significance of the random position of a minor planet circling a wan star in a rather small galaxy, and as a result astrologers selling horoscopes and advice wax while the gullible wane and wait for the stars and planets to control their destinies.

But despite the rise in astrologers, Jesus freaks, drug users, and other harbingers of nonthought, our future cannot be read in crystal balls. "Our destiny lies not in the stars, but in ourselves," wrote Shakespeare taking the side of science in direct opposition to the astrological belief in fate. The world must look to science and engineering for solutions to pollution and other environmental problems. It is true that science has caused today's environmental problems-in the same sense that eating sometimes causes a bellyache. But an occasional bellyache is hardly reason to give up eating. We could eliminate most of our air and water pollution tomorrow by doing away with all cars, using no electricity, and applying no fertilizers or pesticides to crops; in a year, however, most Americans would probably be dead of starvation, disease or exposure. Social modernization coupled with technology can provide realistic solutions to pollution problems. To assert otherwise is dangerous nonsense.

In the short run-the next ten years-pollution should diminish, because the various environmental protection agencies appear to have money and the legal authority to assure the implementation of proven pollution control technologies. Over a hundred environmental protection laws are on the books; most of them are being enforced at least half-heartedly. The decrease in air pollution over most of the nation, the cleaning up of San Diego and San Francisco Bays, the improvement of waterways and estuaries across the land, and the progress made in expanding parks and disposing of solid wastes, all of these obvious gains during the past four years attest to the progress that has been made and the improvements well within reach.

Higher control standards for industrial plants, stronger air and water quality laws, regular progress reports by polluters, penalties and rigid enforcement, all of these obvious techniques work only with sufficient public support. The technology exists for controlling air and water pollution. In January 1971, Ruckelshaus said, "We have known how to control smoke for years...by applying present technology, industry, apartment buildings, and commercial buildings could cut the 17.5 million tons of smoke and soot particles by 95 percent to about 700,000 tons." In a similar vein, Ruckelshaus said, "Untreated municipal and industrial wastes pour into rivers and lakes in every region, although basic treatment technology has been available for a generation." Pollution control will cost more than $200 billion using available technology. Spread over 10 years, $200 billion is a small fraction of the U.S. gross national product, and is a minor percentage of the monies the federal government is likely to waste on nonproductive expenditures. After the 1972 election, the Nixon administration indicated that it intended to spend much less money during 1973 and 1974 on pollution control programs. President Nixon cut deeply into the $11 billion authorized by Congress for water treatment plants and directed that only $5 billion be spent. Air pollution control and solid waste management programs are also slated for reduction. Though progress may not be as rapid as I and many others would have liked, we can still look forward to a less polluted atmosphere during the 1970's.

While considerable progress has been made in cleaning up the environment on an ad hoc basis, society must take a number of positive, conscious long-range steps in order to keep the earth habitable for centuries. Fed by an almost insatiable demand for western-style progress, the pressure for growth at any price threatens eventually to eat up the earth's limited resources. World metal consumption is rising at a rate of six percent per year, which will double the use of most metals in 12 years. Most of the demand comes from that developed one-quarter of the world, the "rich" nations where the annual income is $3,000 or more per person, compared with the other 2.5 billion inhabitants of the globe who have an annual income averaging less than $250 each.

Cheap, abundant energy is the prime requirement of any industrialized society. Modern man's demand for energy has been increasing at a phenomenal rate. During the past 2,000 years, mankind used a total of 12 units of energy, of which four units were consumed during the 19th century. But present consumption is at the rate of 50 units per century, and this rate is doubling every 14 years.

It is obvious that we must provide more energy, that eventually it must come from nuclear fission or fusion. We must also moderate our use of energy, but this alone may not be sufficient to change the world's overall pattern of energy growth.

Technologists are actively investigating a host of new techniques for dealing with present and future environmental problems. Science progresses only by experiment. Better and cleaner systems will evolve only by trial and error. If science has taught mankind anything, it is this: trust only in experimental results. The following list is a sampling of new and largely untried environmental experiments that should go a long way in making the next 30 years cleaner and more pleasant than the past 30 years.

—Creation of islands built of waste, to be used for airports, power plants, and recreational parks.
—Hydroponic farming and aquaculture (growing of fish and shellfish in controlled environments). Much has been done in fish farming; the potential is enormous.
—Long outfall sewer pipes, to simultaneously disperse thermal and biodegradable wastes and fertilize the ocean.
—Advanced high temperature incineration of solid wastes, with recovery of waste heat.
—Conversion of municipal waste into alcohols and useful gases by pyralosis and destructive distillation.
—Recycling of solid waste into building material.
—Reuse of wastewater for plants or for drinking.
—Conversion of oils and vegetable matter directly into protein by bacteria and fungi.
—Use of hydrogen fuel systems.
—More efficient techniques for trapping geothermal and solar energy.
—Fusion energy cources.
—Generation of electric power away from cities, on platforms in the ocean.

SOME CONCLUSIONS

Since the life expectancy of Americans has climbed from 50 years in 1900 to over 70 years in 1971, there is no question that we live in a healthier environment than that enjoyed by our grandfathers despite the fact that our air, water and food have undesirable contaminants in them. Many so-called environmental problems are really popular myths. Our water supplies are good; our normal air supply is not hazardous to health, and air quality has been noticeably improving in most major cities; the solid waste problem is manageable, using present technology such as compacting, land fill, modern incineration and island building; the hazards of DDT, lead, mercury and many other substances have been grossly exaggerated by a sensation-seeking press and a public eager for villains; the ocean is in good shape and likely to remain that way; even most wild animals and trees are doing well and are increasing in numbers.

Yet all animals, including man, would like a "better" environment. Some obvious ways to improve the environment of contemporary America are to replace sprawled suburbs by compact high-rise cities, cut down on packaging, reduce the role of the automobile by providing adequate mass transit, and force industry and government to control pollutants at their sources.

The critical environmental issues in the 1970's will involve modification of contemporary life styles to allow for more rational uses of energy and resources. Life will be centered increasingly in cities and enhancement of city environments is a vital process in full swing today in America, as cars are banned and mass transit planned for dozens of metropolitan areas. Throughout the man-industrialized world large numbers of peasants are moving to the cities. Stress is inevitable, but wise use of high-rise buildings, nuclear energy, and modern transportation and recreational facilities are making our cities more attractive.

Societies of people, like groups of ants or weeds, grow and expand as fast as they can when environmental conditions are favorable. But like ants or weeds, they may, by sheer growth, use up the nutrients and the energy readily available and then die off. For man, the species, the apparant solution to the explosive growth trends in the 20th century is to establish environmental equilibrium: balancing the energy and nutrients that he uses with the nutrients and energy that he can recycle, less those he can obtain for nothing, such as sunlight. The solid, liquid, gaseous and thermal energy that man uses most match the solid, liquid, gaseous and thermal wastes he generates.

A basic psychological question is: why are large numbers of Americans in such an uproar over the environment? Certainly environmental issues are less important than survival issues such as jobs for people who need work; peace, instead of shameful and devastating wars; accomodation between Blacks, Reds and Whites; the survival of our cities. I have an uneasy feeling that environmental issues have predominated because a "sensitive minority" has forced its whims on an unthinking majority. But the ecology issue has been enthusiastically taken up by many people because it appears to be a simple issue with obvious, visible, easily understandable effects. Ecology issues are pleasant to work with because they have a set of obvious villains, "the polluters" (at least when this book was written) and clear-cut solutions which appear to hurt or at least curb the villains. Oh that the problems of the city, of peace and war, of the economy, or of race were so beautifully simple.

FALLING WATERMELONS

We live in mortal danger of being struck dead by watermelons inadvertently dropped from passing airplanes. Technology is obviously to blame for this hazard to life, because, if technologists had not developed airplanes, we would not have to worry about the danger of falling watermelons. However, when we examine the probability that a person will be struck by a falling watermelon, we must conclude that this environmental danger is not worth worrying about.

It is not easy to compute the probability of rare and unlikely events, such as death from eating mercury in tuna fish, cancers in human beings due to DDT ingestion, or the likelihood that polychlorinated biphenyls will have a deleterious effect on the environment. But many people prefer to conclude that new methods, and new substances "may" be harmful and therefore should be banned until-- if ever--they are proved innocuous.

The belief of many that the environmental impact of new chemicals and processes be studied before being released into the environment is commendable in theory. But more than a million products now result from man's activities. There are at least two million species of fauna and flora on the earth. Assuming that a team of biologists can assess the environmental impact of a man-made product on a species in one year's time, it would require some two trillion biologist-person years to complete this worthy evaluation. Such an approach would solve the unemployment problem among biologists for some time to come, but the cost to society would be staggering. Jones W. Haun, of General Mills, Inc., estimates that it costs his company approximately $75,000 for a 12-year program to merely prove the innocuousness to humans of one food additive. Long-term, low-level toxicity studies would cost many times more.

New ideas and new products are delicate things. One consequence of unrestrained precautionary study of possible environmental hazards of new ideas will be to kill most of them before they are born. No child, no idea is born into this world without the possibility of causing harm as it grows older. Shall we then abort all birth and all innovation for fear of possible environmental damage? That is what some environmentalists would have us do, but such a tack would lead society down the road to sterility.

Civilization, that varicolored flower of evolution, has resulted in worldwide conglomerations of large cities and increasingly efficient farms linked together by ships, trains, cars and planes. This is the ecosystem of man in the last third of the 20th century, and it appears stable except for devastating eruptions of native or international war. No one can predict the future: one can merely indicate what man may do to try to insure the survival of his kind.

I began this book with a discussion of myths, and I would like to end with a quotation that touches on myths, by one of the brightest men of the 20th century;

There are certain things that our age needs, and certain things that it should avoid. It needs compassion and a wish that mankind should be happy; it needs the desire for knowledge and the determination to avoid pleasant myths; it needs above all courageous hope and the impulse to creativeness.

The things that it must avoid and that have brought it to the brink of catastrophe are cruelty, envy, greed, competitiveness, search for irrational subjective certainty, and what Freudians call the death wish.

Bertrand Russell, The Impact of Science on Society

The prospect of catastrophe is nothing new to human societies: the Black Plague killed 75 million people and obliterated entire inhabited areas during the 14th Century; the famine of 1878 killed 22 million peasants in China; the atom bomb dropped on Hiroshima wiped out 200,000 lives in a few hours; destruction of topsoil and forests by overgrazing and overfarming and overcutting have caused thousands of societies and hundreds of millions of people to starve or migrate. Only when disease, famine and war are balanced against rational progress and growth, has man lived tolerably well.

From now on man will live in a rapidly changing world in which technology brings comforts and longer life to more and more people. But it will be a world always under the cloud of great potential accidents such as nuclear holocausts, new virulent epidemics, and unpredictable global weather changes. It will be a tolerable world, in many ways more pleasant than the world lived in by our fathers. In the world of the next 200 years, governments will have to work together to manage social institutions and natural resources. And those few persons given to thought will continue thinking while the vast majority of mankind will continue to live intuitively-as has always been the case. It should be a world worth living in.

Ah what a foolish world Allah created,
Ah - what a foolish God Allah must be.
So often in error, so seldom elated,
And great joy is neighbor to bleak misery.

Beauty and bilge,
Black sheep and beasts
Snuggle together
Through famine and feasts.

Why did He do it?
He must have been stoned:
Six days He labored;
He should have first phoned.

Beauty and bilge,
Black sheep and beasts
Snuggle together
Through famine and feasts.

Lord what a foolish world Allah created,
Lord what a foolish God Allah must be.

Peter Agnos

Chapter 1 - CLEARING THE AIR

There are a number of good text books on the science of ecology. I believe that the most widely
used college text is E. P. Odum's Fundamentals of
Ecology (W. B. Saunders Co., Philadelphia, 1971),
a 574 page book crammed with information about the
behavior of various species living in an assort-
ment of ecosystems. Though somewhat cloying, another useful text is R. H. Wagner's Environment
and Man (W. W. Norton and Co., New York, 1970).
Conservation of Natural Resources (edited by
Guy-Harold Smith, J. Wiley and Sons, 1965) provides a well-rounded report on soil, forests, water, mineral and other resources. Anyone who
wishes the negative, pessimistic and paranoid
view of future resources should peruse (it is impossible to read) Limits to Growth, the Club of
Rome computerized report (Universe Books, New
York, 1972) on a dismal future.
Another good general book on the subject is
America's Changing Environment, edited by Roger
Revelle and Hans. L. Landsberg (Houghton Mifflin,
Boston 1970).

Chapter 2 - EVERYBODY & NOBODY POLLUTES

Alexander, Tom, 1972. "No Hysteria About Food Additives," Fortune, March 1972.

American Chemical Society, 1969. Cleaning Our Environment, Washington, D. C., 1969.

American Chemical Society, 1969. Symposium on
Natural Food Toxicants, nineteen papers from the
Atlantic City Symposium in 1968, reprinted from
Journal of Agricultural and Food Chemistry.

Beychok, M. R., 1967. Aqueous Wastes from Petroleum and Petrochemical Plants, John Wiley & Sons, New York.

Bodin, F. and C. F. Cheinesse, 1970. Poisons, McGraw-Hill, New York.

Council on Environmental Quality, 1972 Annual Report.

Esposito, John C. (ed.), 1970. Vanishing Air (A Ralph Nader Study Group Report on Air Pollution), Grossman Publishers, New York.

Fortune, July 1971. Transportation Issue.

Frisken, W. R., 1971. "Extending Industrial Revolution and Climate Change," EOS, 52-7, pp. 500-508.

Kellogg, W. W. and R. D. Cadle, E. R. Allen, A.L. Lazarus, and E. A. Martell, 1972. "The Sulphur Cycle," Science, February 11, 1972.

Kermode, G. O., 1972. "Food Additives," Scientific American, March 1972.

Knapp, C. E., 1970. "Agriculture Poses Waste Problems," Environmental Science and Technology, December 1970, pp. 1098-1100.

Neilands, J. B., G. H. Orlans, E. W. Pfeiffer, Alje Vennema and Arthur H. Westing, 1972. Harvest of Death: Chemical Warfare in Vietnam and Cambodia, Free Press.

Smithsonian Institution, 1970. Center for Short Lived Phenomena, Cambridge, Mass.

"What on Earth is Pollution?", 1970. Interview with Dr. W. T. Pecora in Industry Week, August 17, 1970.

Wolman, Abel, 1971. Talk at the 5th International Water Quality Symposium.

Chapter 3 - POLLUTION WARS & SOCIOECONOMICS
           OF ECOLOGY

Council on Environmental Quality, 1971. Second
Annual Report, Corporate Advertising and the Environment, August 1971.

Houthakker, H. H. (Member, Council of Economic
Advisors), talk before Cleveland Business Economists Club, April 19, 1971.

McGraw Hill, Department of Economics, Pollution
Cost Estimates, 1971.

Neckritz, A. F. and L. B. Ordower, "Ecological
Pornography and the Mass Media," Ecological Law
Quarterly, Spring 1971, pp. 374-399.

Nixon and the Environment, Village Voice Books,
New York, 1972.

Sherrod, H. F., Jr. (ed.), Environment Law Review 1971, Sage Hill Publishers, New York 1972.

The economic impact of pollution control costs
has been slight according to the latest data from
the Environmental Protection Agency. In February
1973, the EPA reported that only 29 industrial
establishments had been shut down during the nearly two years ending in September 1972 in which
environmental costs were mentioned as even a small
contributing factor. The shutdowns involved about
7000 jobs--a miniscule amount compared to the total labor force in the United States.

Chapter 4 - CITY AIR IS KILLING US

Air Resources Board, a joint report to the legis-
lature on air pollution, health effects and emergency actions, Sacramento, California, November
1970.

Auerbach, Oscar, C. Hammond, L. Garfunkel and C.
Benante, "Relations of Smoking and Age to Emphy-
zema," New England Journal of Medicine, April 20,
1972, pp. 853-857.

Auerbach, Oscar, Letter of June 29, 1972.

Battigelli, M. C., 1968. "Sulfur Dioxide and Acute Effects of Air Pollution," Journal of Occupational Medicine, Vol. 10, pp. 500-511.

Beebe, R. G., 1967. "Changes in Visibility Restrictions Over a 20 Year Period," Bulletin of the American Meteorological Society, Vol. 48, p. 348.

Brodie, F. J., 1905. "Decrease of Fog in London During Recent Years," Quarterly Journal of the Royal Meteorological Society, Vol. 31, 15-27.

Chandler, T.J., 1965. The Climate of London, Hutchinson, London.

Council on Environmental Quality (CEQ), 1972. Environmental Quality Annual Report.

Elsaesser, Hugh, "Air Pollution: Our Ecological Alarm and Blessing in Disguise," EOS, March 1971.

Environmental Science and Technology, "Lead", June 6, 1969.

Esposito, J. G., 1970. Vanishing Air--A Nader Report. (A compendium of horrors caused by air pollution)

Freeman, M. H., 1968. "Visibility Statistics for London/Heathrow Airport," Meteorology Magazine, 97, 214-218.

Goldsmith, J. R., 1968. "Effects of Air Pollution on Human Health," in Air Pollution, 2nd Ed., edited by A. C. Stern, vol. 1, p. 547, Academic Press, New York.

Kehoe, R. A., 1969. "Toxicological Appraisal of Lead in Relation to the Tolerable Concentration in the Ambient Air," Journal of the Air Pollution Control Association, vol. 19, pp. 690-700.

Lave, L. B. and E. P. Seskin, "Air Pollution and Human Health," Science, pp. 723-733, August 21, 1970.

Lemke, E. E., G. Thomas and W. E. Zwiacher (eds.), Profile of Air Pollution Control in Los Angeles County, L. A. County Air Pollution Control District, Los Angeles, 1969.

Nadler, A. A., et al, "Air Pollution," Scientists Institute for Public Information Workbook, 1970.

New York City Environmental Protection Administration Reports, "The N. Y. C. Air Pollution Index," 1971.

Oschsner, Alton, "The Health Menace of Tobacco," American Scientist, March, 1971.

Swinnerton, J. W., R. A. Lamontagne and V. J. Linnenbom, "Carbon Monoxide in Rainwater," Science, vol. 172, pp. 943-945, 1971.

U. S. Senate Subcommittee on Pollution, Hearings, 1968, Part 2, pp. 609.

Wimmer, D. B., "Discussion of 'Reactivities of Smog Components are Central Issue in Setting Control Standards'," Environmental Science and Technology, vol. 3, 1969.

Peculiar sociopolitical hearings and gyrations in Los Angeles during March 1973 reiterated the well known facts that 90 percent of Los Angeles fog comes from the six million cars that infest the 9,200 square mile basin. The obvious solution to the problem is to build a comprehensive non-polluting transportation system. There is no likelihood that Los Angeles will meet the 1975 Federal oxidant limit of 0.08 parts per million, a level that is now being exceeded about 200 days each year. Suggestions that gasoline be rationed and car traffic curtailed have been met by howls of protest from automobile-related businesses.

Chapter 5 - THE MYTH OF VANISHING OXYGEN

Broecker, W. S., "Man's Oxygen Reserves," Science, 168, pp. 1537-1538, 1970.

Machta, L. and E. Hughes, "Atmospheric Oxygen in 1967 to 1970," Science, 168, pp.1582-1584, 1970.

Toffler, A., Future Shock, Random House, New York 1970, p. 430.

Wurster, C. F., "DDT Reduces Photosynthesis by Marine Phytoplankton," Science, 159, 1474, 1968.

The relative quiescence of the adherents of ecological doom on the subject of atmospheric oxygen depletion would seem to indicate that some environmental myths can be laid to rest by scientific reasonableness.

Chapter 6 - THE DEATH OF LAKE ERIE

Beeton, A. M., "Changes in the Environment and Biota of the Great Lakes," see Rohlich, "Eutrophication".

Dubos, R., So Human an Animal, 1968.

Ehrlich, P., The Population Bomb, Ballantine Books, Inc., New York 1968.

Hansen, R. S., Great Lakes Dredging, 1971.

International Association on Great Lakes Research, Proceedings of the 12th Conference, Report 1969.

International Joint Commission, Pollution of Lake Erie, Ontario and the International Section of the St. Lawrence River, Report, 1970.

Parker, Carl E., Letter of November 9, 1971.

Rohlich, G. A. (ed.), "Eutrophication: Causes, Consequences, Correctives," Proceedings of a symposium, National Academy of Science, Washington, D. C. 1969.

U. S. Federal Water Pollution Control Administration, Lake Erie Report, August 1968.

U. S. Department of Health Education and Welfare, Report on Pollution of Lake Erie and its Tributaries, 3 volumes, July, 1965.

U. S. Department of Interior, Lake Erie Ohio Intake, Water Quality Survey, 1969, June 1970.

Chapter 7 - OIL ON THE SEA

Allen, A. A., R. S. Schlueter and P. G. Mikolaj,
"Natural Oil Seepage at Coal Oil Point, Santa
Barbara, California," Science, November 27, 1970.

American Petroleum Institute, Report on Air and
Water Conservation Expenditures, 1966-1970, 1971.

Blumer, M., Oceans, XV - No. 2, 1969, p. 5.

Blumer, M. "Scientific Aspects of the Oil Spill
Problem," Environmental Affairs, VI, No. 1, 1971.

Cowan, Edward, Oil and Water--The Torrey Canyon
Disaster, Lippincott Co., 1968.

Environmental Protection Agency--Water Quality
Office, Santa Barbara Oil Spill: Short Term Anal-
ysis of Macroplnkton and Fish, Washington D. C.,
February 1971.

Glazier, F. P. and N. E. Somner, Decreasing Relative Demand for Lubricants: National Petroleum
Refineries Association Meeting, Report AM-70-21.

Jones, L. G., et al, "Just How Serious Was the
Santa Barbara Spill?" Ocean Industry, 1969.

Macklin, J. G., "A Compound of the Effect of Application of Crude Petroleum to Marsh Plants."
Unpublished.

Man's Impact on the Global Environment (MIOGE),
M. I. T. Press, Cambridge, Mass., 1970.

National Air Pollution Control Administration,
Nationwide Inventory of Air Pollution Emissions,
Raleigh, N. C., 1970.

Porricelli, J. D., V. F. Keith and R. L. Storch,
"Tankers and the Ecology," Paper presented at the
Annual Meeting of the Society of Naval Architects
and Marine Engineers, November 11-12, 1971.

Potter, Jeffrey, Disaster by Oil, Macmillan, 1973.

Smith, E. J. (ed.), "Torrey Canyon" Pollution and Marine Life, Cambridge University Press, 1968.

Straughan, D. "The Santa Barbara Study," Joint Conference on Prevention and Control of Oil Spills, December 15-17, 1969.

Straughan, D., "The Influence of Oil and Detergents on Recolonization in the Upper Intertidal Zone," Proceedings of Joint Conference on Prevention and Control of Oil Spills, June 15-17, 1971, American Petroleum Institute, Environmental Protection Agency and United States Coast Guard.

Sierra Club Bulletin, Oil Spill! (listing major accidental spills), February 1971.

Zobell, C. E., "Microbial Modification of Crude Oil in the Sea," Proceedings, Joint Conference on Prevention and Control of Oil Spills, sponsored by the American Petroleum Institute and the Federal Water Pollution Control Authority, December 1969. (Copies available from American Petroleum Institute for $4.00.)

Zobell, C. E., "Sources and Biodegradation of Carcinogenic Hydrocarbons," Proceedings of Joint Conference on Prevention and Control of Oil Spills, Washington D. C., June, 1971. (Copies available from American Petroleum Institute for $9.00.)

Zuckerman, Sir Solly , et al, The Torrey Canyon, Her Majesty's Stationery Office, London, 1967.

In September 1972, Lloyd's Register of Shipping reported that 377 ships with a tonnage of 1,030,560 were lost in 1971--the heaviest loss of ships ever reported in time of peace. Lloyd's noted that of the 1971 losses, 32 percent of the tonnage was accounted for by tankers.

Petroleum imports by the United States have risen sharply since 1970. It is estimated that by 1985 the United States will be importing approximately 15 million barrels per day of petroleum products, most of it by sea.

Chapter 8 - FROTH & FOAM IN THE DETERGENT WARS

American Society of Limnology and Oceanography, "Nutrients and Eutrophication--The Limiting Nutrient Controversy, procedures of a symposium held February 10-12 at the University of Wisconsin, Milwaukee, Wisconsin.

Gibson, C. E., "Nutrient Limitation," Journal of the Water Pollution Control Federation, December 1971.

Grundy, R. D., "Strategies for Control of Man-Made Eutrophication," Environmental Science and Technology, December 1971.

Rukeyser, William S., "Fact and Foam in the Row Over Phosphates," Fortune, January 1972.

Ryther, John H., and William M. Dunston, "Nitrogen, Phosphorus, and Eutrophication on the Coastal Marine Environment," Science, March 12, 1971. (Also see Eutrophication references in chapter 6)

Spong, Senator W. B., "Advice to Consumers of Laundry Detergents," Report to U. S. Senate Committee on Commerce, December 31, 1971.

Making much very public noise, Henry Diamond continues to push for the removal of all phosphates from detergents. In February 1973 he urged New York State to adhere to the proposed June 1, 1973 deadline for banning practically all phosphates from household detergents. This despite the well known fact that 3 out of 4 New Yorkers live in areas where elimination of phosphates would not help receiving waters and where substitutes may cause increased environmental damage.

Chapter 9 - CLOSING THE CIRCLE BY DUMPING WASTES
            INTO THE SEA

Bacon, V. W., "The Land Reclamation Project, Metropolitan Sanitation District of Chicago," (Report) Springfield, Illinois, March 6, 1967.

Bonderson, P. H., et al, "Concepts in Open Coastal Disposal of Municipal Wastewaters," paper presented at Water Pollution Control Federation Conference, San Francisco, October 5, 1971.

Buelow, R. W., "Ocean Disposal of Waste Material," Symposium on Atlantic Shelf, Marine Technology Society Conference, Philadelphia, March 1968.

Burd, R. S., "A Study of Sludge Handling and Disposal," Publication WP-20-4, U. S. Department of the Interior, Federal Water Pollution Control Administration, May 1968.

Chen, C. W., "Effects of San Diego's Wastewater Discharge on the Ocean Environment," Journal of the Water Pollution Control Federation, August 1970.

Council on Environmental Quality, Ocean Dumping, U. S. Government Printing Office, Washington, D.C., October 1970.

Commoner, Barry, The Closing Circle, New York: Alfred Knopf, 1971.

Interstate Sanitation Commission, 1971 Report, New York City, January 1972.

Moiseev, P. A., The Living Resources of the World Ocean (translated from the Russian), U. S. Department of Commerce, Washington, D. C., 1971.

National Academy of Science, "Waste Management Concepts of the Coastal Zone," report by the Committee on Oceanography, Washington, D. C., 1970.

Smith, D. D., Statement before the Subcommittee on Oceans and Atmosphere of the Senate Committee on Commerce, April 21, 1971.

Smith, D. D. and R. P. Brown, "Ocean Disposal of Barge Delivered Liquid and Solid Wastes from U. S. Coastal Cities," U.S. Environmental Protection Agency Office of Solid Waste, 1971.

Tsai, C., "Fish and Sewage Pollution," Chesapeake Science, March 1970.

Wilson, Charles, private letter of September 20, 1966 on milorganite.

Water Pollution Control Federation, "Sludge Disposal: A Case of Limited Alternatives," Workshop of Water Pollution Control Federation at Annual Conference, October 1971, reported in Deeds and Data, December 1971.

Jacques Cousteau claimed that the intensity of life in the world's oceans and seas has diminished by more than 30%, probably by an average of 40% and possibly by nearly 50% in the past twenty years, mainly because of pollution, over-fishing, and environmental changes resulting from human intervention, human over-population being the basic reason.

Chapter 10 - A DASH OF LEAD & MERCURY

Aaronson, Terri, "Mercury in the Environment," Environment, May 1971.

Bodin, F. and C. F. Cheinisse, Poisons, McGraw-Hill, New York, 1970.

D'Itri, F. M., Charles Annett and Arlo Fast, "Comparison of Mercury Levels in an Oligotrophic and a Eutrophic Lake," MTS Journal, vol. 5, no.6,1971.

Elwyn, D., "Childhood Lead Poisoning," Scientist and Citizen, April 1968.

Fogerstrom, T. and A. Jernelov, "Formation of Methyl Mercury from Mercuric Sulphide in Aerobic Organic Sediment," Water Research, 1971.

Grant, Neville, "Mercury in Man," Environment, May 1971.

Guinee, V. F., "Lead Poisoning in New York City," New York Academy of Sciences, May 1971.

Hartung, R. and B. D. Dinman (eds.), Environmental Mercury Contamination, Ann Arbor Science Publishers, Inc., Ann Arbor, Michigan, 1972.

Miller, G. E., et al, "Mercury Concentrations in Museum Specimens of Tuna and Swordfish," Science, March 10, 1972.

Nelson, N. (ed.), "Hazards of Mercury," Environmental Research, vol. 4, Academic Press, 1972.

Tanner, J. T., et al, "Mercury Content of Common Foods," Science, September 22, 1972.

U. S. Department of Health, Education and Welfare, Public Health Service, "Statistics and Epidemiology of Lead Poisoning," Cincinnatti, Ohio, February 1972.

Wallace, R. A., W. Fulkerson, W.D. Sleults and W. S. Lyon, "Mercury in the Environment," The Human Element, Oak Ridge National Laboratory, Oak Ridge, Tennessee, January 1971.

Chapter 11 - A WAR BY MAN AGAINST SOME OF HIS
         NATURAL ENEMIES: INSECTS, POLITICIANS
         AND WEEDS

Borgstrom, Georg, Too Many, Collier Books, New York, 1971.

Frazier, B. E., G. Chesters and G. B. Lee, "Apparent Organochlorine Insecticide Contents of Soils Sampled in 1910," Pesticides Monitoring Journal, September 1970.

Gustafson, C. G., "PCB's Prevalent and Persistent," Environmental Science and Technology, October 1970.

Hammond, A. L., Science, January 14, 1972.

Heath, R. G., "Regarding Mallards & DDT," Nature, October 4, 1969.

Hollander, A. (ed.), Chemical Mutagens, Plenum Press, New York, 1971.

Jefferies, D. J., Nature, May 10, 1969.

Jukes, T. H.,* "DDT, Human Health and the Environment," Environmental Affairs, November 1971.

Knipling, E. F., Interview on trends in regulation of pesticides, Environmental Science & Technology, May 1971.

Laws, E.R., Archives of Environmental Health, August 1971.

Mosser, J.L., N.S. Fisher, T.C. Teng and C.F. Wurster, "Polychlorinated Biphynels: Toxicity to Certain Phytoplankters," Science, January 14, 1972.

Muirhead-Thompson, R.C., Pesticides and Freshwater Fauna, Academic Press, New York, 1971.

Spencer, D.E., "Pollution, Pesticides and Perspective," unpublished paper, October 14, 1971.

Vogel, R. and B. Rohrborn (eds.), Chemical Mutagenesis in Mammals and Man, Springer-Verlag, New York, 1970.

Wilson, C.L. and W.H. Mathews (eds.), Man's Impact on the Global Environment, M.I.T. Press, Cambridge, Mass., 1970.

Wurster, C.F., "DDT and the Environment," in Helfreich (ed.), Agenda for Survival, Yale University Press, New Haven, Conn., 1970.

Wurster, C.F., "DDT Reduces Photosynthesis: by Marine Phytoplankton," Science, 159, 1968.

---

* Dr. Jukes, who is a professor of medical physics at the University of California at Berkeley, has been fighting a rather lonely battle to prevent the banning of DDT. The 30-page article describes the intensive campaign to eliminate manufacture, use and export of DDT in the U. S. by some of the more radical environmentalists. In the fall of 1971 Dr. Jukes was bitterly attacked in print by two members of the Environmental Defense Fund after he had published an article in the New York Times on the relative benefits of DDT.

Chapter 12 - THE ALASKAN PIPELINE

Aspin, L., "Why the Trans-Alaska Pipeline Should be Stopped," Sierra Club Bulletin, June 1971.

Brown, Tom, Oil on Ice, Sierra Club Battlebooks, 1972.

Deason, D., "Permafrost Research," Pipe Line Industry, August 1970.

Deasan, D., "Trans-Alaska: World's Most Thorough-ly Engineered Pipe Line," Pipe Line Industry, August 1971.

Chapter 13 - ELECTRICAL ENERGY & THERMAL POLLUTION

Abrahamson, D. E., Environmental Cost of Electric Power, Scientists Institute for Public Information Workbook, St. Louis.

Coutant, C. C., "Thermal Pollution--Biological Effects," Journal of Water Pollution Control, vol. 43, pp. 1292-1334, 1970.

Krenkel, P.A. and F. L. Parker (eds.), Biological Aspects of Thermal Pollution, Vanderbilt University Press, 1969.

Levin, A.A., T.S. Birch, R.E. Hillman, G.E. Raines, "Thermal Discharges from Electric Utilities," En-vironmental Science and Technology, March 1972.

Parker, F.L. and Peter A. Krenkel (eds.), Engin-eering Aspects of Thermal Pollution, Vanderbilt University Press, 1969.

Resources for the Future, Patterns of U. S. Ener-gy Use and How They Have Evolved, Washington D.C.

U. S. Fish and Wildlife Service, Temperature and Fishes, Government Printing Office, Washington D.C.

U.S. Senate, A Study of Water Pollution, staff report to the Committee on Public Works, Washing-ton D. C., 1963.

Chapter 14 - DEATH FROM RADIATION & NUCLEAR ENERGY

American Chemical Society, Radionuclides in the Environment, 1970.

Dieckamp, H., "The Fast Breeder Reactor," EOS, November 1971.

Gumpert, David, "X-Ray Debate," Wall Street Journal, December 23, 1971.

Morgan, K.Z. and J.E. Turner (eds.), Principles of Radiation Protection, John Wiley & Sons, New York 1967.

Morgan, K.Z., "Adequacy of Present Standards of Radiation Exposure," Environmental Affairs, April 1971.

National Academy of Science, Committee on Oceanography, Radioactivity in the Marine Environment, Washington, 1971.

Novick, Sheldon, The Careless Atom, Dell, New York, 1971.

Sagan, L. A., "Human Costs of Nuclear Power," Science, August 11, 1972.

Sternglass, E.J., Low-Level Radiation, Ballantine Books, New York, 1972.

Stewart, Alice and G.W. Kneale, "Radiation Dose Effects in Relation to Obstetric X-Rays and Childhood Cancers," The Lancet, June 6, 1970.

Terril, J.G.,Jr., paper presented at the annual meeting of the American Public Health Association, Chicago, Illinois, October 11, 1971.

Tsivoglou, E.G., "Nuclear Power: The Social Conflict," Environmental Science and Technology, May 5, 1971.

Chapter 15 - AH WILDERNESS

Andrewaryha, H.G. and L.C. Birch, The Distribution and Abundance of Animals, University of Chicago Press, Chicago, 1954.

Coon, N.C., et al,"Causes of Bald Eagle Mortality, 1960-1965," Journal of Wildlife Diseases, January 1970.

Dubos, Rene, So Human an Animal, Scribners, New York, 1968.

Elton, Charles, The Ecology of Animals and Plants, Methuen & Co., London, 1958.

Hoffer, Eric, First Things, Last Things, Harper & Row, New York, 1971.

Laycock, George, The Alien Animals, Natural History Press, Garden City, New York, 1966.

Robinson, Gordon, "Sierra Club Position on Clear Cutting and Forest Management," Sierra Club Bulletin, February 1971.

UNESCO, "Use and Conservation of the Biosphere, Protection of Rare and Endangered Species,"1970, pp. 143-153.

U. S. Department of Interior, Rare and Endangered Fish and Wildlife of the U. S., Resource Publication No. 34, Washington D. C., 1971.

Ziswiler, V., Extinct and Vanishing Animals, Springer-Verlag, New York, 1967.

Chapter 16 - LIFE & POLLUTION

Commission on Population Growth and the American Future, Final Report, The U.S. Government Printing Office. Washington, D.C. 20402, 1972.
        Despite the fact that the Commission ad-

334

mitted that the United States has a declining birth rate, low population density and enormous amounts of open land it concluded that "the time has come for the United States to adopt a deliberate population policy." The Commission appeared completely unaware that the birth rate was about to drop below the replacement level.

Ehrlich, Paul, The Population Bomb, New York, Alfred A. Knopf, 1971.

Meadows, D. H., D. L. Meadows, J. Randers, W. W. Behrens III, The Limits to Growth, New York, Universe Books, 1972.

National Academy of Sciences and National Research Council, Growth of U.S. Population, (a report), 1965.

National Academy of Sciences and National Research Council, Growth of World Population, (a report), 1963.

Office of Science and Technology, Protecting the World Environment in Light of Population Increase, Washington, D.C., U.S. Government Printing Office, 1970.

Tobin, Maurice, "Environmental Working Papers on International Aspects of Environmental Problems," (unpublished), Washington, D.C., 1971
    An excellent account of a variety of pollution problems and Government actions in this and other countries.

Winger, J. G., Outlook for Energy in the United States to 1985, New York, The Chase Manhattan Bank, 1972.

Chapter 17 - SOLUTIONS & NON SOLUTIONS

17A - SOLID WASTE AND RECYCLING

American Chemical Society, Solid Wastes, articles from Environmental Science and Technology, from 1967 to 1971, published in 1971.

American Paper Institute, "Recycling Waste Paper," proceedings, October 16, 1970, Washington, D.C.

Battelle Memorial Institute, "Solid Waste Processing," 1971 report on results of a survey conducted by the Battelle Memorial Institute, Office of Information, Bureau of Solid Waste Management, 5555 Ridge Avenue, Cincinnati, Ohio 45213.

Bellknap, M., "Paper Recycling: a Business Perspective," New York, New York Chamber of Commerce, September 1972.

Kupchik, G. J., "Recycling and Reclamation," talk before the American Medical Association, April 26, 1971.

Loehwing, D. A., "Garbage Market," Barron's, May 17, 1971.

New York Times, "Birds Today, People Tomorrow," Mayor's Council on the Environment, April 18, 1971.

"Reuse and Recycle of Wastes," Third Annual Northeastern Regional Antipollution Conference, July 21 - 23, 1970, Westport, Connecticut, Technomic Publishing Company, 1970.

U. S. Bureau of Mines, "Third Mineral Waste Utilization Symposium" Chicago, Illinois, March 1972.

17B - A WAY TO GO

Council of Environmental Quality, "Annual Report," Washington, D.C., 1972.

Erlich, A., and P., Population, Resources, Environment, San Francisco, W. H. Freeman and Co., 1970.

Gell-Mann, Murray, "How Scientists Can Really Help," Physics Today, May 1971.

Lewin, J. D. "New York City Power and Pollution Ills," Professional Engineer, February 1972.

Meadows, D.H., D.L. Meadows, J. Randers, W.W.
Benrens III, The Limits to Growth, New York,
Universe Books, 1972.

Murdock, W.M. (Ed.), Environment: Resources, Pollution and Society, Stamford, Connecticut, Sinauer
Publishers, 1971.

National Academy of Sciences, Committee on Resources and Man, Rapid Population Growth,
Washington, D.C., U. S. Government Printing Office,
1971.

Russell, B., The Conquest of Happiness, New York,
Mentor Books, 1951.

Thomas, W.L. (Ed.), Man's Role in Changing the
Face of the Earth, (a symposium), Chicago, University of Chicago Press, 1956.

United Nations Educational, Scientific, and Cultural
Organization, Use and Conservation of the Biosphere,
Paris, 1970.

Zelinsky, W., et al. (eds.) Geography and a Crowding World (a symposium), New York, Oxford University Press, 1970.

Index

hunting, 187
hydrocarbons in air, 30, 88
hydroelectric plants, 217

I

incinerators, 297-298
India, 195, 284, 285
Indian Point power plant, 219
Indonesia, 285
industrial pollution costs, 79
industrial wastes, 36-39
insect sterilization, 192
insecticides, biological, 192
insects, 181ff
Interior Dept., 212
International Commission on
        Radiation Pro-
        tection, 229
International Convention Relat-
        ing to Intervention
        on the High Seas
        in Case of Oil
        Pollution, 134
International Joint Commis-
        sion of Canada
        and the U.S., 113,
        116, 324
iron, recycled, 294
Isaak Walton League, 185
Japan, compacting in, 298
Jeffries, D.J.- finches and
        DDT, 186
Johnson, Samuel, 10
Joint Committee on Atomic
        Energy, 215
Jones, L.G., et. al., 127
Jukes, Dr. Thomas H., 65,
        331

K

Kahn, Fazlur-Sears tower, 304

Kalin, Peter-CCA, 93
Kehoe, P.A.-lead, 99
Kellogg, W.W.- sulphur,
        31
Keystone Canyon, Alaska,
        209
Kneale, G.W.-X rays, 236
Knipling, E.F., 192
Koch, Rep. Edward L.,
        249-250
Kretchmer, Jerome, 27,
        298
Kupchik, George-solid
        waste, 296

L

Lake Erie, 24, 111ff
        description of, 112
        diking of, 119
        fish in, 114-116
        pollutants in, 116-117
        uses of, 113-114
Lake Michigan, 221
Lake Superior, 116
Lang, Martin, 154, 155
Law of the Minimum, 141
Laws, Edward R., Jr.-
        DDT and cancer,
        193
lead, 186
        in the air, 88, 91, 98, 178
        -based paint, 174-175
        colic, 174
        levels in the body, 178-80
        in the sea, 178
        tetraethyl (TEL), 173, 178
lead poisoning, 172ff
        deaths from, 176-177
Lewis, Anthony, 287
Lee, G.B., 197
Liebig, 141
light water reactor, 228

343

myths (cont.)
   spread of, 7

N

Nader, Ralph, 5, 60
National Academy of Sciences, 92, 202
National Agricultural Chemical Assoc., 183
National Environmental Policy Act (NEPA), 68, 71
National Environmental Protection Act, 212, 213
National Parks and Conservation Assoc., 247
National Rifle Assoc., 60
"natural" foods and natural poisons, 33
natural springs, pollution of, 29
nature, 27
   balance of, 23
   legislation, 69
   lovers of, 259
   plan of, 15
   and wilderness, 238ff
Nature Conservancy, 60, 191
NBC Educational Enterprises, 111
Neumann, Einar, 117
New Jersey, 283
New School for Social Research, 12, 111
New York Bay, 132
New York Bight, 132, 152ff
New York Chamber of Commerce, 293
New York City (see also Environmental Protection Admin.)

New York City (cont.)
   air quality, 104
   Bureau of Lead Poisoning Control, 175, 176
   energy use, 270-272
   future of, 302ff
   office space, 274
   Park Dept., 290
   solid waste, 291, 294, 296
   trees, 247
   wildlife, 256-259
New York State Council of Environmental Advisors, 60
New York State Society of Professional Engineers, 220, 271
The New York Times, 10, 119, 140, 143, 147, 213, 331
Niigata Prefecture, 167
nitrates and nitrites in food, 33
nitrogen, 140-142
   oxides, 30
Nixon, Richard M., 80, 228, 312
noise, 49
North Hempstead incinerator, 73-75
North Slope, Alaskan, 205
Northern States Power Co., 73
Novick, Robert, 177
nuclear energy, death from, 227ff
nuclear fission power plants, 227
nuclear reactors, 216
nuclear light-water plants, 217